THE LAST BEACH

THE
LAST
BEACH

ORRIN H. PILKEY AND
J. ANDREW G. COOPER

Duke University Press
Durham and London   2014

© 2014 Duke University Press
All rights reserved
Printed in the United States of America on acid-free paper ∞
Designed by Heather Hensley
Typeset in Minion Pro by Tseng Information Systems, Inc.

Library of Congress Cataloging-in-Publication Data
Pilkey, Orrin H., 1934–
The last beach / Orrin H. Pilkey and J. Andrew G. Cooper.
pages cm
Includes bibliographical references and index.
ISBN 978-0-8223-5798-8 (cloth : alk. paper)
ISBN 978-0-8223-5809-1 (pbk. : alk. paper)
1. Beaches—Environmental aspects.
2. Seashore ecology. I. Cooper, J. A. G. II. Title.
GB451.2.P55 2014
363.700914′6—dc23 2014015437

Image credits:
Cover: Seawall toppled by waves at Nantasket Beach,
Boston Globe/Getty Images; p. ii: *North of Beaufort,
South Carolina*, batik on silk, Mary Edna Fraser, 74" × 36";
p. vi: *Great Ocean Road II, Australia*, batik on silk,
Mary Edna Fraser, 41" × 28".

CONTENTS

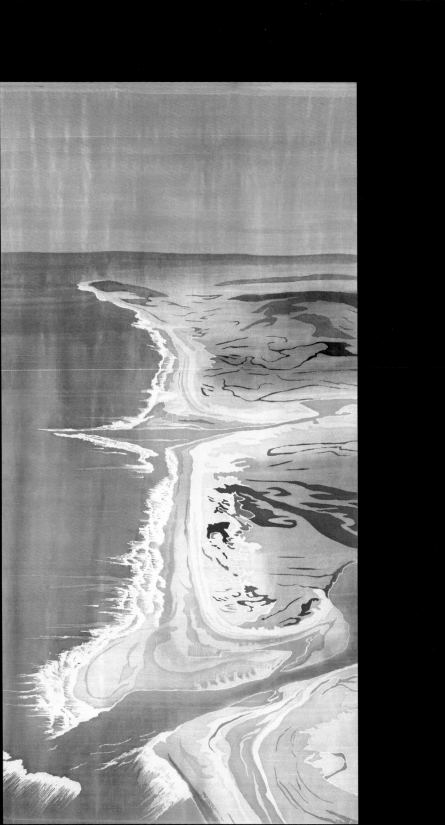

BEACH SCENERY HAS NATURALLY drawn humanity to the seashore.

Serene silences and howling winds, turquoise hues and infuriated white caps, soft sand and rounded pebbles, warmth and cold, recreation and meditation, pain and happiness have guided our steps toward the beach, this unique place where the ever-changing natural elements and human emotions singularly connect.

The beach is the most dynamic and valued feature on the surface of the earth. It is constantly changing shape, absorbing the impacts of storms, and providing a home to an amazing number of living organisms and marine animals.

The beach, sadly, has also become a place of extreme fragility.

Unbridled coastal developments of all kinds—buildings, roads, railways, airports—are threatening and destroying the beaches that attracted us to the shore in the first place.

The widespread damage on western Europe's storm-battered shores, the devastation inflicted by Hurricane Sandy along the northeastern U.S. seaboard, the deaths brought by Typhoon Haiyan in the Philippines—all of these recent tragedies have again exemplified the total inadequacy of infrastructure, and the vulnerability of cities built on the edge of coastlines.

Still, blind to nature's warnings, deaf to its calls, we forcefully rebuild at the very same places, expecting a different result in the name of what is called resilience. Notwithstanding, the seas are rising ineluctably.

The detrimental effects of coastal over-development are heightened by our constant interference in nature. The unabated construction of dams contributes to sand starvation by impeding the natural flow of sediments from mountaintops to rivers, deltas, and finally to beaches. On the coast, seawall construction disrupts the natural movement of sand and waves, hindering the process of sand deposition along the shorelines. Although the beach-destroying impact of seawalls is widely recognized, short-term protection of property has trumped all other concerns.

Less publicized but nonetheless utterly destructive, beach-sand mining gravely exacerbates the chronic demise of the world's beaches.

The beaches, scarred and battered by these inflictions, are furthermore degraded by pollution from many sources such as oil, plastic, chemicals, sewage, and refuse.

Our deference, humility, and respect toward Mother Nature have been lost in the dense fog of our self-perceived invincibility.

Is there hope, or are we to remain blind and deaf to the signs and calls of nature?

As we stand on the beach looking at the present of our feet and gazing at the horizon of our future, we must wonder: Is this to be the last beach? Will generations to come never know the pleasure and beauty of a natural beach?

The loss of beaches will have a detrimental impact on all living beings. The current generation will see it, but it will be our grandchildren and great-grandchildren who will be most affected, first by the soaring costs of defense of the status quo and then by the inevitable beach destruction.

It is not too late, but a major change in perspective is needed in our society.

In this book, Orrin H. Pilkey and J. Andrew G. Cooper advocate a new view of beaches. If we can embrace this ideal, we can yet save the day.

The Santa Aguila Foundation is proud to have made *The Last Beach* possible. The foundation is a U.S. non-profit organization dedicated to the preservation of coastlines around the world, with a focus on global coastal issues and education. Our educational efforts have notably included the management of the website coastalcare.org, the publication of three books co-authored by Orrin H. Pilkey — *The World's Beaches, Global*

Climate Change: A Primer, and *The Last Beach*—and the publication of two books by co-author J. Andrew G. Cooper. We also have brought our support to the prizewinning documentary film *Sand Wars*.

We hope you will enjoy this book about the future of beaches and take as much pride as we do in defending this unique feature of our planet.

Please visit www.coastalcare.org for further information.

THE SANTA AGUILA FOUNDATION

WHAT A PRIVILEGED LIFE WE, the co-authors, have led, visiting, strolling on, and contemplating many of the beaches of the world and studying the complex ways they evolve and change. Most important, we have learned how different types of beaches in different locations respond to rising sea level. Our intuitions about the effect on beaches of sea-level rise, past and present, have been proven by observation: shorelines usually migrate landward but at unpredictable and irregular rates as the sea-level rises. Understanding this, we have observed with alarm the construction of houses, roads, railways, and other urban structures along shorelines around the globe. Two worlds are colliding at the shoreline—the beautiful, flexible, and infinitely adaptable world that is a beach, and the static, inflexible, urban beachfront world.

Long before the present generation, early cultures learned the lessons of living with the shore. The Phoenicians, Greeks, and Romans built ports to facilitate trade and expand their empires. Some of their marine-construction efforts are still visible today, suggesting that they chose harbor sites carefully and with knowledge of the processes of the sea. In general, however, early societies built dwellings and roads a safe distance from the shore. Native Americans visited barrier islands to hunt and fish during summer and retreated off of them when winter storms arrived. Early European residents on North America's barrier islands lived on the back of the islands, following the example of their forebears in Spain and Portugal whose houses in fishing communities faced inland, with their backs to the hostile sea. Of course on a gradually retreating coast, even well-situated dwellings will eventually succumb to the waves. Offshore from the Holderness Coast in Yorkshire, England, the remnants of 28 villages are now under the sea, some as many as three miles from the current shoreline. The villages were abandoned one by one over the 2,000 years since the Roman invasion, as the soft cliffs moved landward. The ongoing cliff erosion provided the sand necessary for the adjacent beaches to survive.

But the rush of civilization to the shore that we have witnessed in our lifetime has been quite unlike anything that occurred in the past. Thousands upon thousands of buildings, some many stories high, are arrayed along beaches that we know are retreating toward them. We know the retreat will continue and will become more rapid as sea-level rise continues. Beachfront buildings sooner or later will be threatened with undermining and will collapse. Anyone could be excused for thinking the world has gone mad. What would a woman from a distant planet say if she came upon a beach like Benidorm or Miami, lined with high-rise development, and knew anything about storms and sea-level rise? She'd be justified in thinking we had gone mad. Yet trying to stop shoreline retreat in the face of rising sea level destroys the beach, and the beach is why we rushed to the shore in the first place and took up residence there. We are killing the goose that laid the golden egg!

As geologists, we are bewildered by this ridiculous and obviously wrong state of affairs, but ours is not to wonder why. Ours is to deal with the present, with the facts on the ground, and to offer a way forward—a way that may yet save our beaches from destruction.

PEOPLE PLACE A VERY HIGH VALUE on beaches. Beaches are an essential part of the coastal landscape everywhere—a central part of local history, admired by visitors and dwellers alike, celebrated in poetry, song, literature, movies, and seascape art.

Like a hypnotic campfire, the breaking surf and wave after wave running up the beach cause otherwise active people to quietly stand and stare. Health professionals have long known the benefits of regular exercise, but when it is associated with a beautiful landscape, they now know that it also improves mental well-being. These values and others are what have made beaches economically important. A worldwide tourism industry is centered on beaches, from the noisy urban beaches of Rio de Janeiro and the French Riviera to the quiet seclusion of Cape Lookout, North Carolina, and the remote beaches of Namibia. Millions of people in coastal countries worldwide derive an income from beach tourism.

Besides its economic attributes, the beach is a wonderful, free, natural defense against the forces of the ocean. Day in and day out, beaches absorb the awesome power of the ocean's waves, reducing them to a gentle swash that laps on the shoreline (that's why there is so much interest in harnessing wave energy for power generation). Even storms don't destroy beaches. Instead, they change their shape and their location, moving sand around to maximize the absorption of wave energy, and then recover in the days, months or even years to follow. Look at those post-hurricane and post-tsunami images of devastation—it's the beachfront houses that are destroyed, but the beaches themselves, albeit covered by debris or covering the ill-placed houses, always remain.

Can we imagine a world with no beaches? As inconceivable as it might seem, such an loss is a distinct possibility, thanks to the way we abuse the shoreline at this time of rising sea level. It isn't that beaches can't handle rising sea levels—they've survived more than 400 feet (130 meters) of sea-level rise since the peak of the last ice age about 20,000 years ago. The problem is the obsession with building and defending property right next to the beach and trying to hold the beach in place. This process, in the long run, destroys the beach. In doing this, we utterly fail to reconcile our love of beaches with our actions; in the end, our contemporary society that favors the preservation of buildings over beaches will be the author of its own destruction. There are various reasons for this; for example, perhaps not enough of us understand beaches, or perhaps the majority who use and value beaches are silent while the few beachfront property owners are more vociferous in demanding protection of their assets.

Many people ask why we don't simply engineer the shoreline and be done with it. This would be a mistake. Trying to hold the shoreline in position makes a flexible response to sea-level rise, one that preserves some semblance of a beach, more difficult. Sea-level rise is anticipated to be minimally 3 feet (1 meter) by the year 2100, a level that will halt most beachfront development on sandy shorelines and barrier islands—unless they are protected on all sides by massive seawalls.

Many beaches have ceased to be the beautiful, self-sustaining ecosystems they once were. Instead they have become long, narrow engineering projects sustained only by constant maintenance and ongoing expenditures. Ugly seawalls, ranging from a few propped-up boards on remote tropical beaches of Colombia and Nigeria to massive towering concrete monoliths on the shores of Japan and Taiwan, have removed beaches altogether. Both private groups and government agencies are pushing the engineering response to beach erosion. The Army Corps of Engineers in the United States, Deltares (formerly Delft Hydraulics) in Holland, the Danish Hydraulic Institute, and HR Wallingford in England (among others), are major movers in shoreline engineering, whose whole raison d'etre is to make money by building defenses. The American Shore and Beach Preservation Association, a strong promoter of beach replenishment, is an example of a private lobbying organization that promotes protection of beachfront buildings. The artificial beaches they advocate are poor imitations of the real thing.

Ironically, beach engineering in its various forms creates a false sense of security that, in the short term, leads to an increase in the density of beachfront development. How else does one explain hundreds of miles of highrise buildings lined up along Florida's largely walled, largely artificial (replenished) beaches or, worse still, the thousands of holiday homes being built and sold on artificially created islands of sand in the Arabian Gulf?

EVEN BEACH REPLENISHMENT, the "soft solution" is only a Band-Aid that must be applied again and again at great cost. It doesn't remove the problem but it treats the symptoms. Eventually and inevitably beach replenishment will stop as either sand or money runs out and rising seas cause erosion rates to substantially increase. Given the political pressure to protect beachfront property, this will inevitably lead to more seawalls which, in time, destroy beaches. Seawalls are the beach's deadliest enemies, preventing their landward migration and causing them to be squeezed out as sea level rises.

Other less obvious forms of engineering have also had an impact on the world's beaches. Dams on rivers that flow to the sea trap sand behind them and starve many beaches, causing them to migrate landward. In California and Portugal the situation is critical and many hare-brained engineered "solutions" have been tried to resolve the situation, but to no avail. Ultimately dams are now being removed to restore the natural system (at least in the Western United States). Deltas are particularly threatened by dams because the loss of muddy sediment causes the land to subside at the same time as beach-building sand is cut off. This causes very high rates of local sea-level rise. Aswan Dam in Egypt has caused beach loss and very rapid erosion of shorelines on the margin of the Nile Delta. The same situation exists on the Niger and Mississippi deltas and is about to happen on the Mekong Delta as upstream dam construction proceeds in China and Laos. One damaging intervention usually leads to another where engineering the coast is concerned; seawalls and groins already are being built along the Nile Delta's beaches.

As if these threats weren't enough, beach mining is a global industry which, in reality, is a direct form of beach erosion. It doesn't take a genius—or even a geologist—to see that taking sand from a beach will make it smaller. Even in the Western world where it is usually prohibited, bucket loads and pickup loads are taken away from many beaches, often in illicit operations known as "midnight supply." The largest beach-sand mining operations at present in the world are on the coasts of Morocco and the countries around Singapore. Morocco mines sand for concrete. In Southeast Asia, massive volumes of beach sand are mined (usually illegally) and exported to Singapore in order to expand its boundaries by infilling nearshore waters and creating new land.

Aside from causing the physical demise of beaches, human activities are also putting many beaches "off limits." Pieces of trash of all shapes, sizes, and compositions end up on the world's beaches. Storms, hurricanes, and tsunamis deposit massive amounts of trash on beaches in short periods of time. Rivers discharge tons of refuse on the shoreline every year. Smaller but more widespread amounts of trash come from passing ships and municipal storm-water runoff.

And what about the water? Each year thousands of beaches are closed to swimmers because of polluted waters. In 2009, almost 19,000 U.S. beaches were either temporarily closed or subject to swimmer advisory days. Storm-water runoff is one of the worst culprits, sweeping up pollutants off the land and washing them out to shores.

Perhaps the most significant factor rendering beaches unusable is polluted beach sands. The fact that beach sand, particularly in the dry beach above normal high tide, is often the most polluted site of all was just recognized in the last decade. Yet beach testing usually involves only the water, not the beach sand. Walking barefoot, lying directly on beach sand, and particularly, being buried in the sand may now be hazardous activities for beachgoers. These are aspects of beach safety that clearly need updating. In the interest of full disclosure, we have a strong personal interest in the beach-pollution problem—co-author Orrin Pilkey's grandson contracted a serious MRSA infection while surfing on a remote beach in Washington state.

On an increasing number of the world's beaches, our foolish and ill-informed actions have meant that the nearshore ecosystem has been lost and that beach fishing, shell collecting, and clam digging are things of the past. The nearshore food chain that originates with the tiny organisms living between grains of sand and surviving on occasional influxes of seaweed is gone. From mole crabs to mackerels, the whole ecosystem is out of whack. Habitats for turtle and bird nesting have been destroyed.

The future of the world's beaches hangs in the balance, from big threats such as engineering, mining, and pollution, to activities that seem harmless, like driving on beaches. The death knell has already sounded for large stretches of beaches along densely developed shorelines like those in Florida, Spain's Costa del Sol, Australia's Gold Coast, and Brazil's Rio de Janeiro. The challenge is to find a new way to live with beaches in a time of rising sea level, reconciling our actions with our love of these beautiful landforms. Given the space, beaches can take care of themselves *and* provide us with multiple benefits. So how do we prevent our beach from becoming the last beach? Our intention in this book is to answer that question—to sound the alarm by giving a fundamental and frank assessment of our current relationship with beaches and the grim future we face if changes are not made. It is not too late, but we must act soon. We advocate a new view, in which we value beaches over buildings, place the enjoyment of the *many* above the self-interest of a *few*, and, in so doing, enable future generations to enjoy beaches in the same ways we have.

ACKNOWLEDGMENTS

FIRST AND FOREMOST, WE GRATEFULLY acknowledge the generous support and continued encouragement from our friends Olaf and Eva Guerrand-Hermes and their Santa Aguila Foundation. Olaf and Eva are friends of beaches and staunch and loyal supporters of efforts to protect them, raising awareness of the issues beaches face via the beach website coastal-care.org, and helping to make books like this one accessible to the widest possible audience.

We greatly appreciate the endorsement of our views and support by the following organizations and individuals: the National Trust in the United Kingdom; Dana Beach, executive director of the Coastal Conservation League; Brent Blackwelder, president emeritus of Friends of the Earth; Pricey Harrison, North Carolina state legislator and a strong supporter of beaches; concerned environmentalist Diane Britz Lotti; and Scott McLucas, president of the One World Foundation, and his wife, Nancy.

We have benefited from discussions with many colleagues over the years. Tonya Clayton, David Fuccillo, Alex Glass, Miles Hayes, Derek Jackson, Joseph Kelley, John McKenna, Bill Neal, and Andy Short have all given us hours of fruitful debate. Bill Neal and Alex Glass also helped with editing comments, some of which were sorely needed. Rob Young and Andy Coburn of the Program for the Study of Developed Shorelines at Western Carolina University provided numerous ideas and data, especially information on the world's beach-replenishment programs. We received a great deal of useful information and leads from the coastalcare.org website and from Claire Le Guern Lytle, the website's manager and a resolute opponent of beach-sand mining. Thanks especially to Norma Longo, able assistant, researcher, copyeditor, and proofreader, whose organizational skills far surpass our joint abilities. Norma was invaluable in keeping things together, sourcing materials, and putting us right.

The publication process with Duke University Press has been smooth and rewarding. Thanks to all at the Press who assisted us with this project, especially our editor, Gisela Fosado, her predecessor, Valerie Millholland, and assistant managing editor, Danielle Szulczewski, as well as editorial associate Lorien Olive, for their encouragement and their patience throughout the process. We truly appreciate the continuous support of our wives, Sharlene and Mandy, and our families who have held down the fort through the months and years we have "endured" the task of visiting, examining, and pondering the fate of the world's beaches.

THE END IS NIGH! 1

ALL OVER THE GLOBE, beaches are moving landward. Where the trappings of humans are absent, the fact that a beach is retreating is neither evident nor of particular concern. But where people have built houses, condominiums, roads, and other structures next to the shoreline, beach retreat becomes beach erosion. This is of great concern to many people, particularly coastal dwellers who value their property. In efforts to hold the shoreline still, today's society is engaged in a costly and ultimately futile battle. On one side is the coastal engineering fraternity and on the other are the inexorable forces of nature. Many beaches on developed coasts have been transformed into long, thin engineering projects on which engineers hold sway until a storm comes along or budgets are squeezed. Ironically, these engineered strips of sand that we call beaches were once a precious natural environment that has been destroyed in a misguided view of the good of humanity.

The hands of humans are very clearly on the beaches of the

FIGURE 1 The end point of hard stabilization on a beach in Albenga, Italy. The groins and seawalls on this beach make it unattractive and dangerous for swimmers—the price paid for holding the shoreline still and protecting buildings. Which is more important, buildings or beaches? PHOTOGRAPH BY ANDREW COOPER.

world. Many of our actions are fairly benign—we swim, fish, sunbathe, stroll, or just enjoy the view, the sea breezes, and the smells of the sea. But we also dump trash and discharge our waste pipes onto beaches. We rake them to "clean them up," drive on them, and mine them for minerals, gravel, and sand. We bulldoze them to make "dunes" to protect houses, pump or truck sand around the beach to "improve" it, and build walls and breakwaters of various types to block waves and hold the sand in place.

NATURAL BEACHES AND HOW THEY WORK

Beaches are things of great beauty. We have only to think of the huge numbers of people who walk on beaches regularly or those who make the long journey to the seaside just to sit and watch the waves breaking, to understand the close relationship between people and beaches. Whether we stare hypnotically at the waves, stroll on the sand looking for seashells, or even brave the waves and get into the water, people all over the world have their own love affairs with beaches. Maybe it is the feeling of nature at work, the thrill of being at the edge of the land, or even the sense of freedom provided by the wide-open spaces.

Maybe it is beaches' dynamic nature that we find fascinating. Why is it that waves can cause cliffs to collapse and push huge boulders around as if they were pebbles, and yet beaches made up of tiny sand grains persist month to month, year to year, century to century? Even the casual observer can't fail to notice that beaches change with time. One day a beach may be steep, a week later, gentle; sometimes there is an abundance of sand, sometimes very little. People often return to a favorite childhood beach to find that it is quite different from what they remember. There are lots of changes to observe and contemplate, and perhaps that is part of the appeal of the beach—it is an ever-changing canvas that is seldom the same from one day to the next. As we will see, it is this ability to change that allows beaches to survive in a hostile environment while solid features, such as cliffs, seawalls, and jetties collapse.

CHANGING BEACHES

Runkerry Strand, on the rugged north coast of Northern Ireland, is a fine sandy beach that is particularly popular among dog walkers in the summer. It is dangerous to swim there, even for anyone prepared to brave the cold waters, because of strong rip currents running directly offshore. Come the winter, however, and walkers are confronted by a beach covered in boulders and pebbles. The rip currents are gone and, instead, the waves break strongly on a newly formed offshore bar. Thanks to the advent of the wet suit, surfers can enjoy the winter waves breaking on the bar.

This complete change in the beach is the result of stronger winter waves, packed with energy from distant storms in the North Atlantic. The sand is unable to withstand the force of the waves and the bottom currents they form, so it is transported offshore, exposing an underlying beach of cobbles (grapefruit-sized rocks) and boulders. When carried offshore, the sand is deposited and molded by the waves into a bar. The bar then helps to break the energy of the waves, and the rest of the energy is

absorbed by the cobble beach. On the other hand, in the summer when waves are smaller, the sand is pushed onshore and is welded to the high part of the beach, covering the cobbles. The waves lose most of their energy by breaking close to shore, and the remaining energy is transformed into rip currents.

A similar winter-summer change occurs on outer Cape Cod beaches in the United States. During the winter, the higher waves form currents that carry beach sand offshore, where three distinct sandbars often are formed. Come summertime, the beaches widen as the smaller waves move sand ashore, causing the sandbars to disappear.

Runkerry and outer Cape Cod beaches, like many others, routinely go through this cycle of sand movement in response to the seasonal wave changes. When it was first recognized by Francis Shepard in the 1950s on the beach in front of the Scripps Institution of Oceanography in La Jolla, California, the two forms of the beach were called *summer* and *winter* profiles. Subsequently, however, it was realized that some beaches changed a lot seasonally while others changed only a little. In the early

FIGURES 2 AND 3 Two examples of beautiful, untouched beaches. On the left is a white sandy beach made up of seashell and coral fragments in the British Virgin Islands, and on the right is a quartz-sand beach in Portsalon, Ireland. There are no erosion problems on beaches such as these until buildings are located next to the shore. As the sea level rises, these beaches will remain, but in a more landward location. PHOTOGRAPHS BY ANDREW COOPER.

1980s, researchers Don Wright of the Virginia Institute of Marine Science and Andy Short of the University of Sydney discovered that the shape of beaches was related to how they absorbed or dissipated the energy carried in the waves. To absorb the energy of large waves, beaches needed to be wide and gently sloping, allowing the energy of the breaking waves to be absorbed over a broad surface. For lesser waves, the energy could be absorbed on a narrow surface as waves swashed up the beach. They termed the two end stages *dissipative* for broad beaches and *reflective* for narrow beaches.

By comparing dozens of beaches in Australia and the United States, Wright and Short found a fairly consistent relationship—the bigger the waves, the larger the volume of the available sand, and the finer the sand, the more dissipative the beach. Although this was an elegant explanation of how many beaches work, no classification made by humans can do justice to a complex natural system like a beach. In fact, it is fair to say that no two beaches operate in the same way.

In 2009, Carlos Loureiro was working on his Ph.D. on beach behavior in southwest Portugal. That winter he noted that the popular surfing beach at Cabanas Velhas lost all its sand. Contrary to expectations (and Wright and Short's theory), however, the sand didn't return the following summer. The reason seemed to be that the winter of 2009 was exceptionally stormy. Storm followed storm and the excess energy in the waves caused rip currents that remained active for long periods, carrying the sand farther offshore than normal. The sand, having been stripped away into deeper water, will take longer to return—and indeed might never return.

It isn't just waves that cause beaches to change shape. Changes may also occur because of a difference in the supply of sand to a beach. Five Finger Strand, a rural beach in County Donegal, Ireland, has seen dramatic changes in the past two decades. This beautiful sandy area near Ireland's remote northern tip began to experience erosion of the beach and its huge, grass-covered dunes in 1995. These dunes were more than 100 feet (30 meters) high and more than a mile (1.6 kilometers) wide. Over the next few years, this erosion continued until the seaward face of the dunes was a 65-foot-high (or about 20 meters) scarp of bare sand and the beach lost its sand entirely, revealing an underlying surface of pebbles and boulders. The local people and county council were concerned by these changes, and researchers at the nearby University of Ulster set out to assess what was happening.

The answer was quite unexpected. The change began when an inlet of an estuary at one end of the beach swung to the north. Normally, the tides and waves at the mouths of inlets build small deltas called tidal deltas. But the small shift to the north was enough to separate the inlet from its delta. Consequently, the forces of the tides and waves began to build a new delta. The sand to build this new delta was drawn in from the adjacent beach and dunes, which ultimately was the cause of the severe erosion of Five Finger Strand. In the meantime, the old delta was freed from the tidal currents that had formed it and held it in place, and it began to be moved by the waves, building up other beaches next to it. So as one beach was being severely eroded because the sand was used to build a

FIGURE 4 A rapidly eroding beach in South Nags Head, North Carolina, with three septic tanks exposed and large sandbags, which offer little protection in storms, under the houses. As the sea level rises, this will be an increasingly common scene, causing beach pollution and destroying any recreational value of beaches. PHOTOGRAPH USED BY PERMISSION OF JOSEPH T. KELLEY.

new tidal delta, a second beach was accreting, using the sand from the old offshore tidal delta.

A study of old maps and air photos shows that this pattern of change had happened before and that the inlet had regularly moved between the two positions. The time frame of shifting was, however, in the region of 25 to 30 years, so this had seemed to be an unprecedented event to most people. But in reality it was just part of a predictable long-term cycle. In 25 to 30 years, the beach at Five Finger Strand should come back. There is some uncertainty about the future, however, because in the last few decades a new element controlling beach evolution has come into play: sea-level rise.

Although the sea has only risen a foot (0.3 meters) over the last 100 years or so, that amount can have a real impact on shoreline retreat on very gently sloping coasts. For example, the average slope of the lower coastal plain of North Carolina is one foot of rise in elevation for every 2,000 feet (610 meters) of horizontal distance. Thus, at least in theory, a 1-foot sea-level rise should push the shoreline back 2,000 feet. On the Outer Banks of North Carolina, the slope is closer to a ratio of 1:10,000, and the shoreline should move back nearly 2 miles (3.2 kilometers). This amount of shoreline retreat will definitely happen, but it will be a delayed action carried out on a multi-decadal time frame. This is because there is a big pile of sand (a barrier island) in the way.

Marine geologist Andy Green was mapping the seabed on South Africa's east coast in the early 2000s when he discovered some remarkable features. Running for tens of miles were low ridges arranged in the same characteristic shapes of a dune line along a modern coastline—but in a water depth of 200 feet (60 meters)! These submerged sand bodies turned out to mark the positions of a former sandy shoreline formed when the sea level was 200 feet below the present and several miles seaward of today's shoreline. The new technology of multibeam bathymetric mapping of the seafloor allows the seabed to be charted in unprecedented detail, which allowed for this amazing discovery.

The sand bodies were preserved as lines of beach rock—a rock created by cemented beach sand that forms on the edge of tropical and subtropical beaches. The dune sands adjacent to the beaches were cemented as well, much as some dune sands in the Bahamas are cemented today. The beaches on the South African seabed were drowned about 11,500 years ago when the sea level jumped 50 feet (15 meters) in just 300 years.

Submerged preserved beaches are now recognized as being quite common—those in the waters off Australia, Brazil, and Florida are as deep as 390 feet (120 meters) below the present sea level. On Florida's Gulf Coast, geologist Al Hine discovered whole sets of easily recognizable forms at 230 feet (70 meters) below sea level. These were once barrier islands, and now they are visible to the multibeam mapper, preserved in minute detail.

The significance of these preserved shorelines and barrier islands, besides offering proof of sea-level change, is that they must have survived as a surf zone moved over them with the rising sea level. The waves in a surf zone could easily destroy these features even in their cemented state, so the sea-level rise must have been rapid. This means that the sea level at that moment in geologic time must have experienced a sudden jump in the rate of rise. A jump might mean 3 feet (1 meter) in two decades (admittedly an educated guess at best). Such a rapid rise would have been caused by a surge in glacier melting or a failure of glacial dams that suddenly released a very large mass of water.

Radiocarbon dating of the beach rock on these abandoned shorelines confirms what we knew from coral-reef records on the steep slopes around the island of Barbados: the sea level has risen by more than 325 feet (100 meters) over the past 20,000 years as the earth moved out of the last ice age and into the warm period that geologists call an *interglacial*. We live in an interglacial at the moment. In fact, over the last 2 million years, the sea level has risen and fallen numerous times as the world's water resources have switched from the ocean to ice caps and glaciers and back again. The submerged beaches, however, also faithfully preserve the shapes of former coastlines and show us that those ancient beaches were quite similar to their modern equivalents. Beaches in nature are almost indestructible.

But it isn't just cemented beaches that tell us of former shorelines. Sometimes the evidence is less direct. Trawlers in Europe's North Sea and on Georges Bank off New England regularly haul up remains of land animals, many of them now extinct—for example, mastodons, saber-toothed tigers, mammoths, and elk. This shows that what is now the seabed was once land.

About 50 years ago off the East Coast of the United States, geologists discovered that under a thin cover of marine sands there would sometimes be mud, and even peat, with the remains of plants and animals (like oysters) that we associate with salt marshes. Clearly these deposits had formed in coastal marshes right at sea level when it was much lower.

Trawlers in Maine had long been bringing up arrowheads and spearpoints in their nets. These caught the attention of archaeologists who discovered that the ancient tools were trawled from a specific site, which prompted marine geologist Joseph Kelley to wonder why the tools were there. Quickly he discovered a submerged landscape preserved on the seafloor at a depth of about 65 feet (20 meters), where early inhabitants of Maine had lived. There were beaches where those inhabitants had made stone tools, fished, and collected shellfish.

Collectively, all of these investigations on the seafloor have enabled geologists to begin to understand what happens to beaches when the sea level rises over thousands of years. Some beaches are stranded and left behind on the seabed, and some roll over the seabed and are reworked into modern beaches, while others are smeared over the seabed, leaving behind a thin layer of sand. Whatever its fate in a rising sea, a beach can usually survive.

Many factors are at work in determining what happens to a beach as the sea level rises. These include the rate and amount of sea level rise, the nature of the beach materials, whether any new material is being added to the beach from rivers and at what rates, where the beach is located (tropics to poles), and the type of beach (e.g., barrier island, mainland beach, pocket beach, or rocky shoreline). The important thing is that beaches have been able to survive more than 325 feet (100 meters) of sea level rise since the last ice age.

STORMS, FLOODS, TSUNAMIS—THE BIG HITTERS

It is one thing to see the evidence of coastal change in the geological record over thousands of years, but we also know that coasts change significantly over much shorter time frames. The passage of a single storm can cause dramatic changes. Countless elderly residents of coastal communities can relate changes on beaches over their lifetimes (changes that sometimes prove to have magnified over time). Many of the most dramatic transformations on beaches occur during big storms, hurricanes, and tsunamis.

The tsunami in the Indian Ocean on Boxing Day in 2004 left many vivid impressions of dramatic changes to beaches. Beaches that had been crowded with tourists were transformed in an instant into a wasteland of debris. In Banda Aceh, the worst-hit coast in Sumatra, geologists recorded the complete loss of beaches and adjacent villages, as the coastline was eroded by more than 325 feet (100 meters) overnight. It seemed as if the coast was utterly destroyed. However, Singapore researcher Soo Chin Liew and colleagues subsequently presented a remarkable set of satellite images that shows the coast before and immediately after the tsunami. The devastation was remarkable, as the beach had disappeared and large swaths of vegetation had been destroyed by the tsunami waves. Yet an image taken in 2006 (only 13 months after the tsunami) shows a newly formed wide sandy beach, admittedly 325 feet (100 meters) landward of the former beach, but nonetheless a coast that hides all vestiges of the recent tragedy, at least on the scale of satellite imagery. From the perspective of the beach, this served to demonstrate its remarkable resilience to devastating waves—the sand that had been on the beach must have been lost offshore only temporarily as it quickly came onshore after the tsunami passed.

Hurricane Sandy (widely known as Superstorm Sandy) struck the New Jersey coast in October 2012. After Hurricane Katrina, it was the second costliest hurricane to strike the United States in terms of damage to property. All along the Jersey shore and its barrier islands, beachfront homes were flooded, the beaches (almost all of which were artificial replenished beaches) were eroded, and several feet of sand were deposited on roads and in formerly neatly tended yards. Within a matter of a few weeks, most of that sand had been bulldozed back to the beach.

Dramatic photographs appeared in the press showing the undermining of a roller coaster at a fairground in Seaside Park, New Jersey. The effects seemed devastating, and people could only stand by and watch this magnificent demonstration of nature's power. What was happening of course was that these beaches were responding to the storm in the way that they always do. Much of the sand that was lost from the beach was the same sand that was being deposited on the roads and yards—the beach was rolling over itself in response to the storm and, had it been a natural beach, when the storm was over it would have looked much as it had before—just repositioned a little farther landward.

Most New Jersey beaches are on barrier islands, and beach evolution is a critical part of the evolution of barrier islands. Ironically, bulldozing sand back to the beach works against the barrier island's ultimate survival. Barrier islands require both storms and rising sea levels in order to survive and evolve. Big storms such as Sandy wash sand over the island and into the bay behind it, a process that widens the island. Much of the overwash sand remains on the island, raising its elevation. Simultaneously, the storm typically moves the shoreline back (which is called shoreline retreat).

Widening the back side, eroding the front side, and raising the interior of the island is the process of island migration. The island moves back and up and thereby responds to sea-level rise. All three steps required for barrier-island evolution are halted on developed islands, and soon the islands become inert piles of sand dependent on costly engineering for their survival.

IN MARCH 2007, THE POPULAR BEACHES of KwaZulu-Natal in South Africa were hit by high waves that coincided with an exceptionally high tide. The tourist beaches around Durban suffered dramatic changes, with sand being carried offshore and sand dunes that had been covered in dense forest being undermined. Five-hundred-year-old shell middens (piles of oyster or clam shells left behind after they had been harvested and shucked by ancient coastal dwellers) were exposed on rocky headlands as the covering sand was stripped away.

The loss of sand from the beaches was big news, and the authorities were very concerned. Durban geologist Alan Smith and his colleagues concluded that the combination of previous storms that winter, which had caused the beaches to narrow, and the big waves on top of an exceptionally high tide were responsible for the unprecedented amount of shoreline erosion. The rare combination of events was estimated to have a chance of repetition only once in five hundred years—a figure supported by the shell middens. Massive rip currents produced by the waves caused much of the sand loss. These currents drove sand into deep water from which it may take years or even decades to return, and some of it may never return. In the meantime, the beaches have shifted landward and are slowly regaining their pre-storm dimensions, but in a more landward position.

Studies by Paul Gayes of Coastal University in South Carolina after Hurricane Hugo (1989), and other studies by Miles Hayes, then a graduate student at the University of Texas, after Hurricane Carla (1961), show extensive offshore transport of beach sand, probably beyond the middle of the continental shelf. This sand is not likely to return to the shallow water.

WHERE DOES THE SAND COME FROM?

It is important to understand how beaches change, but an even more fundamental question is: where does the sand (or cobbles or boulders) come from in the first place? Related to this is the question of whether sand is still entering the beach or is being lost from it.

FIGURES 5, 6, AND 7
Three crowded beaches
narrowed by seawalls.
With time, these beaches
will likely disappear
unless new sand is added.
The first is in Esterel,
France (PHOTOGRAPH BY
ANDREW COOPER); the
second is in Turunç,
Turkey (PHOTOGRAPH BY
ANDREW COOPER); the
third is Portavecchia
Beach in Monopoli, Italy
(PHOTOGRAPH USED BY
PERMISSION OF NORMA
LONGO).

Waves can work with whatever material is available to build a beach, and most beach sands come from multiple sources. On steep mountainous coasts, such as the west coast of all of North and South America, and on river deltas, the rivers directly supply sand to the beaches. On flat coastal-plain coasts, such as the eastern United States, Brazil, China, and Mozambique, rivers dump their sand far from the beach at the heads of estuaries. The sand on today's beaches is largely material that was deposited there when the heads of the estuaries were located where the beaches are. That was at a time when the sea level was lower. Also, a portion of the sand in many beaches traveled up the continental shelf, pushed by the waves, as the sea level rose.

All beaches of the world get some of their sand by lateral transport by waves in what are called longshore currents. This sand may be from adjacent beaches or from nearby eroding bluffs and cliffs. Geologist Robert Morton demonstrated on a Texas beach that sand blown to the beach from dunes, especially after storms, was an important sand source for some beaches.

We now recognize that sand being trapped behind dams on rivers is a major cause of beach erosion. This is especially true on mountainous coasts such as the west coast of North America or the south coast of Spain. Beaches adjacent to coastal plains are generally not affected by dams, because the beaches there are not directly supplied by today's rivers. Instead, the river-borne sand is deposited far inland at the heads of the estuaries where river currents stop and sand drops out. Recognition of the problem caused by dams has led to the removal of many small dams in the United States, with more and larger ones likely to be removed in the future. The 108-foot-high (33-meter) Elwha Dam in Olympic National Park, Washington, was built in 1910 and blown up in September 2011, by which time the dam had collected about 6 million cubic yards of sediment and it was of limited use for power generation. Before the dam was built, the river had supplied fine-grained beach sand, which provided a suitable habitat for the clam population important to the Elwha Klallam Native American tribe that inhabited the Elwha Delta. After the dam was built, the fine sand no longer came down the river, and with no new sand arriving, the breaking waves on the beach took away the smaller grains, leaving only gravel. As a consequence, the clams disappeared. Two years after the destruction of the dam, both the fine sand and the clams were beginning to return and the beaches were becoming wider.

Erosion of bluffs and unconsolidated glacial deposits is another important source of beach sand. All along the southeast coast of England are beaches of flint cobbles. The flints are eroded from the famous white-chalk cliffs of Dover (the chalk breaks down and is washed away, but the hard flints remain), and continuing erosion of the cliffs is necessary for the survival of the beaches. Unfortunately, on many shorelines around the world, cliff erosion has been halted by efforts to protect cliff-top property by building seawalls up against the bluffs. As a consequence, beaches are being starved of sediment (just as dams starve beaches) and are eroding. The same is true in some areas of New England, where beaches depend on the erosion of glacial deposits for their survival, but the deposits are now "protected" by seawalls and erosion has been halted. Had the link between cliff erosion and beach building been understood, perhaps things would have worked out differently: the supply of sand from the eroding bluffs would continue and beaches would continue to flourish.

Other major sources of beach-building material are the shells and skeletons of marine creatures. Many of the clams, snails, sea urchins, sponges, and microscopic sea creatures that have hard parts in their bodies contribute to some extent to beach building. It is not uncommon in the tropics to find beaches adjacent to coral reefs that are composed entirely of shell and coral remains. Almost all beaches contain at least a trace of shelly material. As some shells become broken down, others are being produced, so these beaches can generate a self-sustaining sand supply.

A strange source of sand is found near Fort Bragg, California, on what is now known as Glass Beach. For many years, starting in the early twentieth century, local residents threw their bottles and other garbage onto the beach. The result, after decades of wave action, is a colorful beach of sand-sized rounded glass fragments. Another unique beach, near Liverpool, England, has an unusual red color derived from disposed brick fragments from buildings destroyed by bombing during World War II. On England's Durham Coast, coal waste was dumped on the beach for decades. The beaches are being restored at massive cost and still have sections that are so acidic that nothing can live in them.

In all these cases, it is easy to see that the rates at which sand is added to or removed from the beach change with time. If more is lost than is added, a beach is likely to move landward, and, conversely, if more is added than lost, the beach will advance seaward, as with the beaches near the mouth of the Columbia River in Washington, or at Whatipu Beach near Auckland, New Zealand. Some beaches are advancing seaward for this reason, but estimates indicate that 90 percent of the world's beaches are retreating.

LIFE IN THE BEACH

Aside from humans, many other creatures also appreciate the beach as a place to reside, and still more visit beaches at important stages in their lives. These creatures interact with the beach and the surrounding water to create an ecosystem, a self-contained and distinctive system that is every bit as complex as the better-known coral-reef or rainforest ecosystems.

Most of the inhabitants of the beach are all but invisible to us. Sometimes a hole in the sand or an unusual trail on the surface gives a clue as to what lives there, but apart from the activities of crabs and shorebirds, much of the action is underground. Inhabitants of the beach have to be able to survive in its constantly moving environment. Many burrow in the sand to avoid drying out at low tide, to avoid being washed away, or for protection from predators.

Scientists divide the plants and animals in an ecosystem in many different ways, but from a practical perspective, perhaps the most common division among the animal inhabitants of sandy beaches is according to size: macrofauna and meiofauna.

Macrofauna refers to the biggest creatures: mollusks (e.g., clams and snails), crustaceans (e.g., crabs and shrimps, and small creatures known as isopods and amphipods), and polychaetes (marine worms) in the beach ecosystem. They usually eat smaller creatures or filter food suspended in the water and are themselves eaten by seabirds, surf-zone fish, and mammals. Most of these creatures are able to burrow rapidly in the sand for protection.

Meiofauna is the name given to the smaller organisms that live between the sand grains on the beach. Many are barely visible to the naked eye, but they occur in huge numbers (up to millions per square foot of beach). The meiofauna form a link between the microbial community and the macrofauna. They also break down detritus in the beach and make it available to other creatures. At the same time, they help keep beaches clean. They are very sensitive to environmental disturbance, and when a beach is polluted, they are quickly replaced by nematodes (tiny worm-like creatures).

The smallest of all are the bacteria, nematodes, flatworms, and some smaller amphipods and copepods. Some of these creatures live in the water, while others live on the surface of individual sand grains. These creatures generally play a role in the breakdown of organic matter, providing nutrients for other living things.

To understand beaches as ecosystems, we have to think of a series of relationships between animals, plants (algae), and their environment. Some of these relationships involve one creature eating another, while others are more mutually beneficial—creatures removing nutrients and keeping water suitable for other creatures. The basis of the beach as an ecosystem is energy in the form of food, most of which is transported to the beach from the sea and deposited there. In a few beaches, algae (microscopic, planktonic plants that grow in the sea) form the base of the food web, but on most beaches, the basic food is material thrown onto the beach by waves (beachcast), such as seaweed, sea grass, and floating natural debris. These basic ingredients sustain the whole beach ecosystem. Some of it is eaten directly by birds, insects, and crabs scavenging on the dry parts of the beach and by shrimp and fish on its submerged parts. The feeding action is often concentrated along the drift line, where the latest deposit of food has been washed up. Foraging birds can often be seen running along this line, avoiding the swash as it runs up the beach.

Debris such as seaweed carried to the beach forms a physical obstacle to the wind and traps windblown sand, leading to the development of dunes. The seaweed provides nutrients for the initial growth of dune plants. Plant growth can also benefit from the waste products of birds that roost on the beach.

Aside from their permanent residents, beaches can be vital to temporary visitors. Sea turtles, in particular, are strongly reliant on beaches. Although they spend most of their life at sea, turtles must visit beaches to lay their eggs. They usually have favored beaches to which they return time and again. They drag their heavy bodies up the beach under the cover of darkness, dig a deep hole with their hind flippers, and then lay hundreds of eggs before filling in the hole and returning to the sea. When the baby turtles hatch they become part of the beach ecosystem. Even before they hatch, foxes, raccoons, crabs, birds, and even dogs dig up and eat turtle eggs. On the beaches of Ras al Hadd in Oman, turtle eggs make up the main part (up to 95 percent) of the diet of Arabian foxes.

Around the world, many unexpected creatures depend on the beach ecosystem in one way or another. In Namibia the lions have adapted to life in this hostile desert environment by scavenging on dead whale carcasses and other edible remains washed up on the beach. In Gabon the tropical rainforest reaches the sea in the Loango National Park, bringing the most unexpected sight of hippos, buffalos, and elephants wandering across the beach and swimming in the sea. Although these creatures do not depend on the beach for their survival, it has been speculated that the hippos bathe in the seawater to remove parasites.

Many commercial species of fish and shellfish rely on beaches for some or all of their life cycle. In northern Europe, the juveniles of several species of flatfish (for example, Dover sole, plaice, and flounder) wander the sandy intertidal parts of beaches, feeding on crustaceans, polychaetes, and the siphons of bivalves that protrude above the sandy surface, as they in turn filter food from the water column. Annual commercial catches of Dover sole in northwest Europe are in the region of 35,000 tons. In the United Kingdom alone, 3,700 tons of plaice were landed in 2010 with a first-sale value of £21.7 million. At a much smaller but no-less-important scale, many rural coastal communities around the world manage to earn a livelihood from collecting shellfish living in sandy beaches, much as people have done for centuries.

When the Maputaland Marine Reserve was first established in South Africa in 1986, a study was undertaken by a local conservation scientist, Robert (Scotty) Kyle, to determine whether subsistence harvesting of ghost crabs and mole crabs from the beach was sustainable. His study showed that the harvesters caught an average of twenty to thirty crabs per night and that most were eaten by the immediate family. He concluded that the practice was not endangering local stocks.

Many species of fish, including commercially important ones, rely on food from the beach ecosystem. When humans interfere as we so often do, it is not just the beach that we must consider. What about all the creatures in the ecosystem? We could be damaging fish stocks, endangering turtles, or polluting the water. Whether driving on beaches (killing crabs

FIGURE 8 An elephant seal resting on Drake's Beach at Point Reyes National Seashore, California. Obviously, the bluff at the back of the beach is a source of sand, and when such bluffs are seawalled to protect buildings on the top, the sand supply is lost and erosion accelerates. PHOTOGRAPH BY ANDREW COOPER.

and trapping juvenile turtles), building seawalls (reducing the amount of beach), scraping beaches (denuding them of their nutrient sources and killing almost everything), or engineering beaches (altering the amount and type of sand), the potential for humans to damage the beach ecosystem is vast.

FIGURE 9 King penguins at Volunteer Point, a headland on the east coast of East Falkland in the Falkland Islands. This beautiful beach has no buildings that hinder shoreline retreat, and these are the most common creatures seen on the beach. PHOTOGRAPH USED BY PERMISSION OF JOSEPH T. KELLEY.

THE BEAUTY OF BEACHES

Beaches exist in many different settings, change in different ways, and are sustained by sediment from different sources, but in all cases they are mobile and able to change shape in response to changing conditions—flexibility is the reason for their success.

It is the job of a coastal geologist to understand what makes a particular beach or group of beaches exist and change shape, and in so doing to better understand these wonderful landforms. It should be clear by now that no two beaches are alike. We can make some generalizations and group beaches according to common characteristics, but there is no one-size-fits-all scheme for understanding beaches.

What they have in common is their ability to change as conditions do. The fact that beaches transform—from day to day, year to year, decade to decade, in response to storms or groups of storms, variations in sediment input, and even sea-level fluctuations over thousands of years—is the secret of their success.

Storms, calm weather, changing sea levels, floods, waves, sand, and gravel all are ingredients in the beach recipe. We often read that the sea level poses a threat to beaches, or that storms threaten beaches; such a view is absolutely untrue of natural beaches. Beaches have survived and even benefited from countless storms and more than 325 feet (100 meters) of sea-level rise during the 18,000 years since the close of the last ice age, without the help of any human hand.

It is often difficult to convey to the general public the beauty of the beach and the great value of the beach ecosystem in such a way as to raise people's awareness and appreciation of the impacts of our actions on beaches. This is true of many different types of environments, and one way that has emerged recently is to think of the benefits that different ecosystems bring to humans. This has been brought together in the concept of ecosystem services, which identifies the benefits of natural ecosystems to humans and then attempts to assign an economic value to these. The first part is relatively straightforward while the second part is fraught with difficulty. The list below outlines the main ecosystem services provided by beaches:

- *Storm protection*: dunes and wide beaches protect buildings from storms.
- *Nutrient cycling*: beach animals are important to both marine and terrestrial food webs as they cycle nutrients.
- *Biological cleaning*: beaches are large biological filters; the sand and the organisms that live in it clean and detoxify coastal waters through a process known as biofiltration.
- *Nesting and breeding sites*: beaches function as food stores and nursery grounds for many organisms (e.g., turtles, birds, and fish).
- *Recreation and tourism*: beaches are tourism hot spots and important to local economies.
- *Fishing (both commercial and recreational)*: some animals (clams and crabs) are harvested for food and bait.
- *Commercial fishing*: many commercially important species of fish and shellfish depend on beaches.

- *Conservation*: rare and endangered species, including turtles and birds, spend crucial parts of their life cycle on beaches.
- *Health*: outdoor recreation improves people's health and the thousands of beach walkers no doubt receive some such health benefit, even setting aside the mental benefits of enjoyment of the beach.

It is easy to see when things are laid out like this that beaches bring us many different types of advantages, and we can start to envisage the economic value of each one. In the beginning beaches are simply malleable piles of sand, but in the end they are much, much more than that.

A good analogy for the beach environment is life itself. Living things are systems that operate smoothly and evolve in predictable and sensible ways to enhance their survival and health, as are beaches. Viewing beaches as living things provides a basis of how to live with them in a way that will keep them alive and evolving, a way to preserve them for future generations. The difference between life and death as it applies to a beach should provide guidance to distinguish good and bad development practices. Some of the beach's life processes follow:

Beaches require sustenance. Sand is food to beaches.

Beaches use energy. Waves, and to a lesser extent wind, tides, plants, and animals, furnish energy to the beaches.

Beaches grow fat. A large sand supply or a stable sea level results in island or beach widening.

Beaches grow thin. Sand starvation or rapid sea-level rise may thin them down.

Beaches protect themselves. Beaches flatten to cause the energy of waves to be dissipated over a wide area.

Beaches recover. Sand comes back to beaches and dunes after the big storms, and the beach may return to a steeper, narrower state.

Beaches have friends and enemies. Storms are their friends. Engineers who arrest their evolution are their enemies.

Beaches have different personalities. Different combinations of sand supply, beach material, sea-level change, subsidence, vegetation type, orientation, oceanographic setting, and wave climate are responsible for large differences in the nature of beaches.

Beaches don't die. They can be killed by people but not by nature.

In the following chapters we will demonstrate that the actions of humans all around the globe are, unfortunately, leading to the degradation and widespread loss of beaches. With the exception of the highly visible fauna of beaches, such as nesting turtles and birds, the beach ecosystem is almost never a consideration when evaluating the impact of human activities on beaches.

STRUCTURE OF THE BOOK

In the rest of this book, we consider two categories of issues that threaten beaches as we know them. One category is concerned with human activities that damage and degrade beaches. Driving, pollution, oil, and trash each have the potential to diminish beaches, reducing their attractiveness, popularity, and utility, as well as damaging their ecosystems. When this happens, beaches begin to lose the political support necessary for their preservation in a time of rising sea levels. The second category is concerned with threats to the very existence of beaches. These activities—hard and soft engineering plus beach mining—are destroying beaches throughout the developed world. Through globalization, all of these practices are being exported worldwide. We explore the reasons why and expose the murky world of coastal engineering.

2 SELLING THE FAMILY SILVER

Beach-Sand Mining

ONE CERTAIN WAY OF DESTROYING a beach is to dig up and remove some or all of its sand. In spite of this obvious link, taking sand from beaches and dunes is a global phenomenon that has taken place for as long as humans have lived near the shore. Initially, it must have occurred on the scale of a basket- or bucketful, graduating to horse-drawn carts that were loaded manually, and finally to dump trucks serviced by mechanical diggers. Now it occurs on some beaches at the scale of long lines of large dump trucks each containing as much as 13 cubic yards (10 cubic meters) of sand. And it's not just sand. Pebbles, gravel, and even boulders are taken from beaches for various purposes. Most beaches can survive small sediment losses, but persistent sand removal over a long time, or massive one-time removals, can prove fatal.

The continuous activity of surf-zone waves usually produces a uniform (well-sorted) sand size that is free of mud. This means that beaches are often seen as ideal sources of

FIGURE 10 Mining of the beach sand trapped by the jetty north of the mouth of the Ria de Aveiro, Portugal. PHOTOGRAPH USED BY PERMISSION OF WILLIAM J. NEAL. THIS PHOTOGRAPH ORIGINALLY APPEARED IN WILLIAM J. NEAL AND ORRIN H. PILKEY'S ARTICLE "BEACH MINING: ECONOMIC DEVELOPMENT / ENVIRONMENTAL CRISIS."

clean sand for making concrete and a variety of other uses in construction. Beach mining is cheap because it is easy to excavate sand with front-end loaders, backhoes, or shovels. Perhaps most significantly, beaches are seldom privately owned; most are public and are perceived as a free source of aggregate to those willing to take it. In regions such as North Africa, where forests are sparse or absent and lumber is a costly import, concrete rather than timber is used for construction. On small islands, beaches are a convenient (and often the only) source of sand for building.

Coastal sand mining has become a major problem as demand rises and humans devise more-efficient ways to remove large volumes of sand. Ironically, some of the demand for such sand comes from beachfront development, which poses a serious threat to the very existence of beaches, the safety of beachfront development, and the future of coastal tourism. The problem is most severe in developing countries that lack the means to regulate sand removal.

In recognition of the damage done by sand extraction, many countries effectively ban large-scale mining. Yet, an article from 2009 by Robert Young and Adam Griffith on the educational website coastalcare.org shows that beaches remain an important sand source in more than 30 countries. It seems that every developing country with a coastal city has a sand-mining problem. Where there is a building boom, there is a mining boom. Beach-sand mining, both legal and illegal, is very widespread on both sides of India. In some areas (especially near Mumbai), a "sand mafia" has evolved, threatening and intimidating opponents, competitors, and the media, and bribing officials to look the other way. Crushed rock has been widely advocated in India as an alternative to both beach and river sand, but to date, production of sand in this fashion has not stemmed the more-profitable but illegal trade in beach sand. An equally strong and dangerous sand mafia exists in northeastern Algeria. There, huge amounts of sand destined mostly for the construction industry are mined on a daily basis along a 6-mile (10-kilometer) stretch of beach east of Jijel.

Some beach and dune mining is for minerals that are in the sand. In the list below, the product sought by miners is in parentheses. The minerals include ilmenite and rutile (titanium), zircon (zirconium), magnetite (iron), garnet (abrasives), cassiterite (tin), monazite (rare earth elements), native gold, chromite, and, in Namibia, diamonds. In some cases, the sand is returned to the beach once the desired mineral has been removed. The impact on beaches from these activities depends on how much material is removed and the care that is taken in restoring beaches and dunes after mining. Although it is opposed by some conservationists, two decades of large-scale dune mining in KwaZulu-Natal in South Africa have been accompanied by intensive dune restoration and replanting of vegetation to reinstate the dune ecology.

Some beach mining is downright thievery. A spectacular example of this was the complete disappearance of a quarter mile (or about 400 meters) of gleaming white-sand beach in Coral Springs, Jamaica. The stolen beach, theft of which was discovered in July of 2008, had a volume estimated at 500 truckloads. It was going to be the grand centerpiece of a proposed new (but now canceled) US$108 million resort hotel. The stolen beach was a major event in Jamaica, and the prime minister, Bruce Golding, personally monitored an extensive investigation which, in the end, led nowhere.

Much sand stealing goes on under the cover of darkness, which was probably the case in Jamaica. In Cancún, Mexico, seven workers were arrested for erecting beach-engineering structures at night in front of the Gran Caribe Real Resort. This happened in spite of the fact that five hotel workers had previously been arrested for the same crime. The situation was brought to light when adjacent hotel owners complained that the Gran Caribe Real Resort was trapping sand and stealing their beaches.

Small-scale sand mining goes on everywhere, particularly on small islands. Throughout the Caribbean, mining of beach sand is assumed to be a reason for significantly narrowed (relative to their reported widths by early settlers) beaches on most of the islands. In Puerto Rico, beach mining has recently been spotted going on within yards of signs stating that beach-sand mining is illegal. We have seen bathtub-sized depressions in beach or dune sand at many locations on the U.S. mainland and in Hawaii. These were likely due to small pickup truck loads, obtained after dark, and destined for use on small nearby construction projects. In Washington state and in Nome, Alaska, small-scale panning or sluice-box gold mining is allowed on beaches, a process that disturbs the sand, affecting the beach fauna, but it does not remove significant volumes of sand from the beach. In the British Isles, there is a long history of sand removal from beaches for agricultural purposes. The poor glacial soils of north and western regions were improved by the addition of seaweed and sand (containing shells). The practice is now illegal in most places but still occurs. In the Outer Hebrides of Scotland, many farmers will eventually concede that they have taken sand from beaches (and still do). In the 1960s there was even a government-sponsored scheme to increase the amount of arable land by mixing beach sand into peat bogs, and almost every building on the islands is built on a base of beach sand.

In Northern Ireland the scenic beach of Cushendun in the Glens of Antrim is owned by the National Trust, a conservation charity in the United Kingdom. There was, however, a long history of farmers removing sand from this beach. In the past the sand was manually shoveled and removed by horse and cart, but recently tractors were used to dig up large quantities. Calculations suggested that at current rates of extraction the beach only had another 50 years or so before it would be mined out. The site already had a fast shoreline-retreat rate, and historic buildings would be threatened in the near future. The farmers insisted that they had a traditional (and possibly legally defensible) right to remove sand and would be exercising that right in spite of the consequences for the beach. In the end, the beach mining was stopped only when the National Trust bought glacial sand and placed it in a parking lot adjacent to the beach so that farmers could freely take sand from that pile.

Just a few miles along the coast from Cushendun, an enterprising but crooked individual with a holiday home near the shore tunneled from his house and under a public road to get to a small pebble beach. He mined the beach illegally and sold the pebbles (used for ornamental purposes) to garden centers. By the time he was discovered and stopped, much of the beach had disappeared.

There is little legal mining of beaches in the United States for construction sand (concrete). One rather strange exception is the 200,000 cubic yards (153,000 cubic meters) of sand mined annually at the CEMEX plant along the Monterey Bay, California, shoreline. The sand is actually extracted from a pond next to the beach but located above the mean high-tide line. The pond is filled every winter by storm waves that wash sand from the beach into the pond. The U.S. Army Corps of Engineers halted other Monterey Beach mining operations but could not touch the CEMEX mine because it is above the high-tide line, beyond their jurisdiction.

In KwaZulu-Natal, the numerous small-but-steep rivers deliver much of the sand to the region's beaches on which its thriving tourism industry is based. Still, dozens of "sand-winning" operations are licensed on the rivers. These operations remove large quantities of sand before it even reaches the beaches and pose a threat to beach survival. Dredging activity may create a number of negative environmental impacts, including release of toxic chemicals from bottom sediments. This can include heavy metals such as mercury and compounds such as PCBs, DDT, and other organic toxins, which are then ingested by the local fauna. The mud stirred up by most of these operations can kill some organisms, especially filter feeders, and can silt-up nearby marshes and mangroves. Another well-documented long-term effect of river mining is a downstream change in the shape and location of meanders of the river channel, creating erosion problems. A different problem was created in Goa, India, where river-mouth mining allowed the penetration of high-salinity ocean waters much farther upstream than before the mining began.

A similar effect can be created by mining sand offshore and thus starving the beach. The country that moves about the largest volume of near-shore sand in the world is the United States, with its beach-replenishment program. Since about 1965, at least 370 million cubic yards (283 million cubic meters) of sand have been pumped onto beaches, today costing at least US$3.7 billion. (This is a process discussed in detail in chapter 4.) Most of this sand has come from the continental shelf offshore of the new artificial beach, sometimes as far as 20 miles (32 kilometers), but usually much closer. This process is devastating to both beach and continental shelf fauna and flora and may change the nature of the waves striking the shoreline, increasing the rates of shoreline retreat. Flocks of seagulls, feasting on the bodies of sea creatures killed by the dredge, mark the site of most in-progress replenishment projects.

Mining the sand from San Francisco Bay has been going on for more than a hundred years. First, the mining was part of the deepening of the bay for the port, but more recently it has been carried out to obtain construction sand. It is now clear that the loss of sand has had an impact on local beaches, in particular the southern portion of Ocean Beach, which is one of the most rapidly eroding beaches in the entire state of California. There is increasing pressure to halt mining in the bay and to increase the amount of sand imported from British Columbia, where it is obtained from deposits left behind by glaciers. In higher latitudes, sand deposited by former glaciers is an important construction sand source, which lowers the pressure on beaches.

Sometimes beaches are even mined to provide sand for other beaches. After Hurricane Hugo in 1989, the bulbous ends of some barrier islands in South Carolina were mined to provide sand for central eroded portions of the islands, a perfect example of robbing Peter to pay Paul.

In the 1930s, sand imported from Manhattan Beach, California, was placed on Waikiki Beach in Hawaii. Eventually, sand from an Australian beach also made its way to Waikiki. More recently, sand from other Hawaiian beaches has been used on Waikiki. All this sand dumped on one of the most popular beaches of the world has had unexpected consequences. The sand flowing offshore from the beach during storms apparently is responsible for killing a large section of the fringing coral reef. The result has been a gap in the formerly protective reef, which has led to higher waves and more-rapid erosion of the replenished beaches, which then require more sand, creating a troublesome cycle.

FOUR EXAMPLES OF "SERIOUS" BEACH MINING
Singapore

According to *Asian Beat*, in 2013 Singapore's national development minister, Tan Chuan-Jin, stated that Malaysian beach sand "is good for nothing—lying about on the beach all day. Now the grains are 'pulling their weight' by being part of Singapore's glorious expansion. Each grain can now have pride in itself with a decent work ethic." According to the development minister, the lazy Malaysian sand is clearly better off working in Singapore! The United Kingdom's *Daily Telegraph* presented a different view: "Singapore, the island city-state, has been accused of launching a clandestine 'Sand War' against its neighbors by paying smugglers to steal entire beaches under the cover of night."

Sand has been a critical element in the extraordinary economic success of Singapore, with its society that prides itself on environmentally sound and sustainable development. The economic success of this former British colony has made it one of the four Asian Tigers, along with South Korea, Taiwan, and Hong Kong. It is a highly urbanized society (5.3 million residents) with little space left for a growing population. The city-state's areal extent has increased about 20 percent since the 1960s, from 224 to 272 square miles (580 to 705 square kilometers). An additional 20 percent of seaward expansion is planned to be completed by the year 2030. The sand for this land expansion, called land reclamation, will be mined from the nearshore as well as from beaches, islands, and rivers—exclusively from neighboring countries.

Much of the sand arrives in Singapore by what is sometimes referred to as "midnight supply." Smugglers dash in at night, remove the beaches of Indonesia, Cambodia, and Malaysia, load the sand into small barges, and sail straight for Singapore, where apparently no questions are asked. The border islands off the north coast of Indonesia, which lie in close proximity to Singapore, offer a particularly tempting target for sand thieves. Some tiny Indonesian islands will likely disappear as a result of the mining. One interesting ramification of island disappearance could be a reduction in the size of Indonesia's territorial waters.

Singapore imports around 15 million tons of sand per year. Cambodia is the most important sand source, with 800,000 tons alone coming from the nearby Koh Kong Province. Next, in order of volume contributed to Singapore, are Indonesia, Malaysia, Burma, and the Philippines.

In 1997 Malaysia imposed a ban on sand export, which was followed by similar bans in Indonesia (2007), Cambodia (2009), and Vietnam (2012). Unfortunately, the region is rife with corruption, and the bans have apparently done little to slow the flow of sand. For example, 700 highly visible truckloads of marine sand cross the bridge to Singapore from Malaysia every day. Once, when a rumor spread that the country was about to enforce its rules about beach sand, a traffic jam occurred on the highway leading to Singapore as drivers fearing arrest abandoned hundreds of trucks. Much of the sand involved comes directly from beaches, ruining any potential for future beach tourism. Other sand is mined from rivers, depriving beaches of a future sand supply that would have come down to the beaches in floods.

Obtaining sand from dredging rivers creates problems exemplified by the experience of the ecotourism village of Tatai, Cambodia, near the mouth of the Tatai River. Extensive around-the-clock mining of the river began in 2012, causing immediate loss of the local fishing industry and the disappearance of tourists. Although some of the dredgers denied that the sand was for Singapore (one company claimed that it was simply deepening the navigation channel), investigation showed that Singapore was indeed the sand's destination.

In terms of its impact on sand mining, Singapore has become a regional pariah responsible for an economic and environmental catastrophe for which the island city-state has avoided responsibility. At present, Singapore's Building and Construction Authority says that the source of the sand is not public information. The Ministry of National Development maintains that the sand is purchased from approved sources. Hopefully, Singapore officials will get their heads out of the sand and start ensuring that future sand will be mined from environmentally acceptable locations and paid for in full.

Morocco

This North African desert kingdom is a constitutional monarchy with a population of 32 million. It has coasts on both the Mediterranean Sea and the North Atlantic and is a land with very wide beaches and miles and

FIGURE 11 Illegal sand mining of dunes and beaches on a massive scale in Morocco.
PHOTOGRAPH © SANTA AGUILA FOUNDATION, COASTALCARE.ORG.

miles of coastal dunes that are among the largest in the world. These huge volumes of sand were probably derived from deserts that once occupied the continental shelf when the sea level was low. Each time the sea level rose, over the past 2 million years, waves pushed more sand onshore.

Morocco is also a land with no significant source of wood for construction. In order to make up for the lack of low-cost lumber, Morocco mines its beaches and dunes to make concrete. The amount of sand mined annually is probably equivalent to the volume of sand brought into Singapore in a year. The big difference is that in Morocco, the extracted beach sand is being used in the same country from which it was mined, while Singapore is taking sand from other nations. Whether the sand is derived locally or from surrounding countries, in either case, beaches are being destroyed.

In July 2007 we observed an operation south of Rabat that involved the

FIGURE 12 Piles of sand brought up from a Moroccan beach by donkeys. The sand is then loaded into trucks for distribution elsewhere. PHOTOGRAPH BY ORRIN PILKEY.

removal of dune sand by front-end loaders filling up hundreds of dump trucks each day. The demand is such that on the beaches, dump trucks are often loaded by the shovelful when larger equipment is not available. In some cases, beach sand is even carried up bluffs on the backs of donkeys in long trains from beaches not accessible by trucks. The damage to the beaches, particularly in northern Morocco, is stunning. When the miners finally leave a Moroccan beach-dune mining site, a lunar-like landscape remains.

Aside from creating dramatic changes in scenery, the impacts of this sand mining extend to the beach (threatening fauna and flora and nesting shorebirds and sea turtles), the dunes (damaging the endemic and sometimes-rare vegetation), and the coastal wetlands (disturbing migratory waterfowl, among other organisms). Removing sand from the beach and from adjacent dunes increases the erosion rate for the impacted shoreline at the same time as the rising sea level is increasing the threat of long-term coastal erosion. Neighboring, un-mined shorelines are likely to experience an increase in erosion as the shoreline adjusts. In addition, the protection from storms, tsunamis, and other large wave events that is provided by wide beaches and large coastal dunes is being removed. Mining of beaches and especially dunes increases the vulnerability of all coastal infrastructure and ecosystems that were once protected by the sands.

What is the solution for Morocco? Is there some way the country can satisfy the voracious sand appetite of contractors without damaging the country's precious beaches? The answer would appear to lie in the deserts of Morocco. Over the coastal Atlas Mountains are thousands of square miles of desert, much of it covered by dunes. Tapping this huge source of sand would probably require construction of a railroad or highway to the coast or over the mountains. While mining inevitably results in some environmental changes, removing sand from the desert is preferable to removing it from beaches, which damages tourism potential and the protective nature of beaches.

There is a widespread perception that desert-dune sand is too spherical to be useful in high-strength concrete. Sphericity of sand grains is a measure of how close a grain comes to being a perfect sphere. No data from widespread deserts are available to support this concern, and since desert sand comes from many sources, it is likely that at least some of it is suitable for concrete. Finding a desert-sand source for construction will likely require prospecting to find the properly shaped grains. We believe that the alleged unsuitability of desert sand may be an urban legend.

Sierra Leone

Sierra Leone is a tiny country on the west coast of Africa bordered by Guinea to the north and the west and Liberia to the south. The country is famous for its pristine, palm-tree-lined beaches near the capital city of Freetown, which received its name when it was declared a home for former slaves freed from British colonies. Between 1991 and 2002 Sierra Leone experienced a devastating civil war in which fifty thousand people died, and 2 million out of a population of 6 million became refugees in surrounding countries. The nation's infrastructure and many of its buildings were destroyed.

Now a huge reconstruction effort is under way. It is not surprising that the reconstruction is consuming vast amounts of beach sand. Unfortunately, compared to Morocco, the natural volume of sand on Sierra Leone's beaches is miniscule, and there is a real possibility of completely destroying any future beach-oriented tourist industry in a few years.

According to Kolleh Bangura, the director of the Sierra Leone Environmental Protection Agency, the rate of sand removal is accelerating, and the shoreline-erosion rate has increased sixfold in beach-mining areas. It is regrettable that local chiefs and councils have the final authority over sand mining, and local chiefs are said to be getting rich from the flourishing industry of sand mining. The Environmental Protection Agency is powerless.

Starting in 2012 on Hamilton Beach, five miles from Freetown, as many as 40 trucks could be seen lined up daily, each with its own crew of laborers armed with shovels. As one beach is destroyed, the trucks move on to the next. The youths of Sierra Leone have an unemployment rate of 70 percent, and beach mining is labor intensive. The need for postwar employment is one reason why the government is having a difficult time halting beach mining. Providing alternative employment for local youths would be a big step toward reducing beach destruction.

Today Hamilton Beach is closed to mining—and for that matter closed to anything else. Sand extraction exposed the bedrock and coarse gravel, and the pre-war tourist beach is no more. Shoreline-retreat rates are now very high, and a number of buildings will soon fall in, including an orphanage run by OrphFund. According to a 2013 estimate, the orphanage has two years to go before tumbling into the sea. The beach on Hamilton Beach is gone, the tourists are gone, and the jobs are gone.

John Obey Beach, 20 miles (32 kilometers) south of Freetown, is the site of a new ecotourism project by Tribewanted, a sustainable ecotourism organization. The organization is in its fourth year of building cottages, digging wells, constructing toilets, and attracting foreign tourists. But now this thriving ecotourism destination has been designated as a mining site. Even though the sand miners have agreed to stay away from the beach immediately in front of the ecotourism community, the removal of sand from adjacent beaches is causing the loss of sand in front of the ecotourism site and threatening the burgeoning tourist economy.

The obvious solution for Sierra Leone is to find another source of sand, but no other source will be as cheap as beach sand while there is some left. Alternatives include importing sand, mining rivers far upstream from the shoreline, and crushing rock to form sand. Before any of these alternatives can happen, the political and employment situation must be resolved and national control of beaches must be restored, which is a tall order for any country and perhaps an impossibility for Sierra Leone.

FIGURE 13 Filling a dump truck with sand on a Sierra Leone beach. Mining by hand pays many otherwise unemployed young men, but the mining is damaging the future of the tourist industry. PHOTOGRAPH USED BY PERMISSION OF TOMMY TRENCHARD / IRIN.

FIGURE 14 Intense mining of a Sierra Leone beach. The sand here is removed down to the bedrock and is then used to make concrete to repair and replace the large number of buildings destroyed in the recent revolution. PHOTOGRAPH USED BY PERMISSION OF TOMMY TRENCHARD / IRIN.

Barbuda

This 160-square-mile (414-square-kilometer) island with a population of just 1,600 is a part of Antigua and Barbuda, a Lesser Antilles country. Antigua, a slightly smaller island of 108 square miles (280 square kilometers), has a much larger population—80,100. Tourism is not as important on Barbuda as on Antigua, but for those who come here, one of Barbuda's favorite tourist spots is the Frigate Bird Sanctuary, where 5,000 frigate birds reside along with 170 other species. Some argue that this extraordinary bird sanctuary should be the basis of an expanded tourist industry.

There are two tourist facilities on Barbuda and another one is planned, but the primary and by far the most visible industry on the island is sand mining. Barbuda is one of the lowest-lying islands in the Caribbean and, except for the wetlands, the entire island was covered by sand dunes made up of sand brought to the beach by waves and then blown inland. Mining has gone on since 1980, and local journalist Rory Butler noted that the last completely natural sand dune was mined in 2012. On our visit to Barbuda 10 years ago, much of the island looked like a war zone.

Mining the dunes is removing the very soul of this tiny island. The high topography that would have afforded some protection for future development in a time of rising sea levels is gone. The beautiful forested rolling topography is gone. Mining continues and lowers the elevation of the island with every truckload of sand that is dumped into a waiting sea-going barge.

Between 1980 and 1997 two million tons of beautiful white sand worth US$300 million were mined by Antigua Aggregates Limited. The sand was sold to a number of island communities in the Caribbean. From 1980 to 1994 the mining operators received $218 million while the island government received $6 million. Now Barbuda receives $40 per ton compared to the $1 per ton it received in 1996. (Actually, the sand tonnage reported by the mining company is suspect because the count was made by the number of truckloads rather than the actual weight of the loads.)

In 1997 a judge halted the sand mining because of environmental concerns and because no money was coming to the people from the mining operation. But in a few years it was back on track, and tons of sand were once again being loaded onto barges for shipment throughout the Caribbean. Among the sand customers are Saint Martin and various islands in the British and U.S. Virgin Islands. On Virgin Gorda, a new development that stresses its green credentials has publicized building lots ranging from $2 million to $25 million and has used 5,000 tons of beautiful white sand from Barbuda to build its beach.

Arthur Nibbs, the chairman of the Barbuda Council, once a strong opponent of mining, managed a complete turnabout, arguing that mining is now essential for the salvation of the island's people. Nibbs now says that his hands are tied and that he can't halt the mining: "Would you prefer to be protecting the environment and then have your people go hungry?"

In Barbuda sand mining takes sand that cannot be replaced except by nature, a process that may take thousands of years. Because Barbuda dune sand is diminishing in volume, it is expected that beach sand will be the next source of sand for mining activities. Some beach sand has already been mined, and there is talk of getting offshore sand, a process expected to kill the coral reefs.

Mining Barbuda is destroying the island's future.

Grenada, another Caribbean island, presents a situation very similar to Barbuda; in this case, the circumstances are highly political. Sand mining in Grenada has had a devastating effect on the island's beaches, which are critical to the tourist industry. In 2009 beach mining was halted by the National Democratic Congress administration, but in 2013 the New National Party administration reversed the anti-mining policy "to boost the island's construction industry."

In 2012 George Worme, the editor of the local Grenada paper *New Today*, was arrested for stealing four buckets of beach sand. It seemed to be a strange arrest on an island where many tons of sand are being removed from local beaches. It has been suggested that the arrest was probably a convenient charge to harass an editor who accused the president of bribery.

WHERE TO NEXT?

In all four cases, sand mining is a lucrative activity from which unscrupulous operators stand to make a lot of money. It is equally clear that the first step in halting the beach-destroying sand-mining industry is, in all cases, political. Putting it another way, sand mining on beaches is always facilitated by bad government.

Mining sand from any beach should be illegal. This is almost always already the case, but regulations are meaningless unless enforced. National governments must lead the way in protecting these valuable national resources. This might even require providing appropriate infrastructure for alternative sources of sand.

In a perfect world, Singaporean sand importers would stop bribing officials of other countries to facilitate the theft of their sand. The other countries that are victims of the thievery must stiffen up their bureaucracies and punish those accepting the bribes. In Morocco the king must declare a halt to beach-sand mining and facilitate opening up the desert sand source. In Sierra Leone, achieving the perfect world is an even bigger challenge. If the national government were to take over control of beaches, they might be able to save them, but some local councils say this would start a war. In Barbuda, mining is considered to be essential to the survival of the island's economy. But what does that mean if the sand mining destroys any possibility of a future tourist economy? The islanders are caught between a rock and a hard place.

Perhaps the tiny Caribbean country of Saint Lucia illustrates the best solution, or at least sets a good example. Saint Lucia is a volcanic island about the size of Singapore, but its population is approximately 1 percent that of Singapore's, and Saint Lucia is not trying to increase its area by land reclamation. Thus, its sand requirements are relatively small, although the country is replenishing some tourist beaches. Saint Lucia has designated a pocket beach enclosed by two large capes as a legal mining site. Removal of sand there is very unlikely to affect adjacent beaches. At the same time, there is a rock-crushing operation on the sides of the volcano to provide sand for construction.

THE BOTTOM LINE

Mining sand from beaches has no positive side effects aside from the fact that it is the cheapest and highest-quality sand available for construction use and beach replenishment in most coastal areas. Trumping that economic advantage is the economic damage that mining does to the tourism industry. In addition, beach-sand mining is environmentally damaging

FIGURE 15 The mining of an eroding gravel beach in northern Sumatra, Indonesia, for construction aggregate. PHOTOGRAPH BY MARIANNE O'CONNOR, FROM ORRIN H. PILKEY, ET AL., *THE WORLD'S BEACHES.*

in the extreme, as it removes natural coastal protection and destroys the beach and nearshore ecosystem as well as any associated fishing activities. Added to all of this is the reality that it is a global source of widespread corruption among those charged with enforcing mining restrictions. Sand mining weakens the resilience of beaches, and this effect is doubled when combined with sea-level rise. In 2013 sand mining garnered international attention because the United Nations included it in a list of new and emerging issues.

INDEFENSIBLE   3

Hard Structures on Soft Sand

A NATURAL BEACH NEVER needs protection from the forces of nature; it is perfectly adapted to cope with anything that is thrown at it. Without the impact of humans, a beach will always be there and in good shape, although after a storm, it might be a different shape or in a new location. Clearly, however, the hands of humans *are* upon the beaches of the world at the same time as the level of the sea is rising (and as we jam our buildings as close to the sea as we can). Bizarre as it may seem in this context, the closer to the sea, the more valuable the property!

The greatest threat to the future of the world's beaches is posed by the coastal-engineering profession. The engineer's primary charges are to protect beachfront buildings and enable navigation in and out of ports. To accomplish this, engineers attempt to hold shorelines still, despite the fact that flexibility is essential to the survival of beaches. It is a Herculean task by any measure, but in the face of occasional storms and rising sea level, such efforts are the equivalent of putting Band-Aids on gaping wounds.

There are three fundamental responses available to our society for shorelines that are retreating in a landward direction: (1) *soft stabilization*, meaning beach replenishment; (2) *hard stabilization*, meaning seawalls; and (3) *retreat* back from the shoreline. The first two are an attempt to hold the beaches in place. The third is an approach that lets nature do the adjusting to climate change.

Shoreline stabilization is engineer-speak for holding a shoreline in place when it doesn't want to be held in place. However the task is carried out, it bodes ill for the beach. Yet shoreline stabilization is a global practice based on a centuries-old societal priority that values beaches less than buildings. The common view has changed little since Sir John Rennie (the Younger) asserted in his 1845 address to the Institute of Civil Engineers: "Where can we find nobler or more elevated pursuits than our own; whether it be to interpose a barrier against the raging ocean . . . ?" In reality, constructing such barriers in a time of rising sea level, when most sandy shorelines are retreating, comes down to an important decision: do you want buildings or beaches? Take your pick. You can't have both.

Coastal engineering, a subspecialty of civil engineering, was first practiced along the shores of the Mediterranean, the Red Sea, and the Persian Gulf, possibly before 3500 BC, and perhaps a little later in China. In those times beaches didn't draw the attention of engineers. Harbor construction and navigation-channel construction and protection were the principal tasks. A great deal of ingenuity was required to build safe harbors and wave shelters without the modern means of moving big rocks and huge volumes of sand that we take for granted today.

The practical know-how of ancient engineering gradually improved through the Phoenician, Greek, and Egyptian societies, reaching its zenith in the days of the Roman Empire. Romans learned to build underwater stone structures with vertical walls, using cement that hardened when submerged, and sometimes the Romans attached stones together with metal bands. At the same time, the Romans developed the technology of deepening navigation channels and ports by dredging. In England, starting with the Roman occupiers, low-lying lands, such as those around the River Medway, were protected by embankments, and elsewhere in Eng-

FIGURE 16 The armored entrance to a harbor in Knidos, Turkey, that was originally constructed by early Greek engineers. In those days there was no mechanical equipment such as cranes or front-end loaders to move the rocks about. PHOTOGRAPH BY ANDREW COOPER.

land and in the Netherlands, engineers reclaimed or protected land by the use of dikes.

As the Western world became more populous and prosperous, beaches beckoned with the promise of relaxation, fresh air, ocean waves, and revitalization. Although the use of beaches for recreation and tourism is usually associated with the Industrial Revolution, it began centuries earlier. The modern town of Frejus on the French Riviera originated as a seaside resort for veterans of the eighth Roman legion in about 30 BC, and Pompeii, Italy, was a popular location among Romans for holidays at the time of the volcanic eruption that destroyed the city in 79 AD. In the United States, the New Jersey shore was the first to develop a large beach-oriented tourist industry, beginning in the early nineteenth century. It

FIGURE 17 This North Sea sand and shingle beach near Lowestoft, England, shows multiple attempts at hard stabilization, most of which have been destroyed, similar to those along the New Jersey shoreline before its burial by beach replenishment. This beach has an additional problem, however, as some of the offshore debris remains from structures built to repel invaders during World War II. Now a popular leisure activity here is promenading on the top of the seawall. PHOTOGRAPH USED BY PERMISSION OF GINA LONGO.

should come as no surprise that the skills that engineers learned in building harbors and reclaiming land would come into play at coastal tourist resorts from an early date.

THE IMPACT OF ENGINEERS ON BEACHES

In many places today, beaches, much like highways and canals, have come to be regarded as long, narrow engineering projects. A highly competitive field of engineering firms, agencies, and consultants offers a range of "solutions" to an ever-increasing number of people whose property abuts ocean shorelines that are retreating landward. These solutions include an array of methods that range from ridiculous and fanciful "alternative" approaches to the brutal and damaging traditional barriers against the sea.

The problems that engineers create for beaches begin far inland with the construction of sand-trapping river dams that starve beaches of sand, thus leading to shoreline erosion. This is a particular problem for rivers that flow directly to the sea and not through large estuaries. Thus, on the west coasts of both North and South America, many beaches are retreating at accelerated rates because of the theft of sand by dams. As part of a growing trend, the 108-foot-high (33-meter) dam on the Elwha River in the Olympic Peninsula of Washington state was recently removed, in part to restore sand flow to the beaches. (The situation is described in chapter 1.)

The complications resulting from dams are most pronounced on the world's deltas, including those of the Mississippi, Nile, Niger, Yangtze, Ganges, and Indus Rivers. Sand mining of riverbeds only adds to the sand-loss problem. The impact of engineers on beaches continues on coasts where dredged navigation channels at inlets and deepened harbors further disrupt sand supplies to sand-starved beaches, and ends with the myriad activities of engineers who try to tame beaches and hold them in place.

Beach engineers are different from other civil engineers who deal in lumber, concrete, and steel. Those who design bridges, water tanks, buildings, and highways have to worry about the safety of humans. They can, with reasonable accuracy, predict the response of their structures to natural processes such as earthquakes, floods, and wind. Of course, building design is also helped by a lot of experience, along with substantial safety factors to cover the unexpected. And a limit is always recognized (such as tornado winds) beyond which fail-safe construction is economically impossible. At least in a relative sense, designing with natural processes in the construction industry is a straightforward matter. If something goes wrong, it is obvious to all.

In his 1951 memoir, the famous engineer (and former U.S. president) Herbert Hoover wrote: "The great liability of the engineer compared to men of other professions is that his works are out in the open where all can see them. His acts, step by step, are in hard substance. He cannot bury his mistakes in the grave like the doctors. He cannot argue them into thin air or blame the judge like the lawyers. He cannot, like the architects, cover his failures with trees and vines. He cannot, like the politicians, screen his shortcomings by blaming his opponents and hope that the people will forget. The engineer simply cannot deny that he did it. If his works do not work, he is damned."

Unfortunately, the same is not normally true of coastal engineers, whose mistakes are blamed on unexpected storms or acts of God and are usually forgiven. Coastal engineers deal with the performance of a dynamic natural system, which is anything but straightforward, and their attitude when things go wrong is quite different from that of engineers like Hoover. Those who engineer beaches do so on the world's most precious and most storm-tossed pieces of real estate. Storms are the biggest problem. They are responsible for most of the changes in beaches, and who among us can predict when the next storm will come, where it will strike, how long it will last, from what direction it will come, how intense it will be, and how frequently others will follow? Rarely are engineers called upon to save the beach itself, although their efforts are often cloaked in the holy shawl of beach preservation. Instead, they are asked to hold the shoreline still and preserve roads and buildings with the impossible expectation that the beach won't be damaged in the process. They are also supposed to inform the public about what will happen to the beach after the engineers have left town. But this is something they simply can't do with reasonable accuracy.

One thing that coastal engineers and other civil engineers and architects have in common is politics. Politics led to the lack of building-code enforcement and to the poor construction of schools in Sichuan Province, China. As a result, thousands of schoolchildren died in the 2008 earthquake. The politics on beaches is often manifested by greatly smoothed-over predictions of environmental impacts of hard structures, concocted in order to satisfy politicians and obtain project approval (and funding).

A 2011 projection based on "sophisticated and state-of-the-art" mathematical models divined that a proposed set of four groins (walls built perpendicular to the shoreline) along the mid-island shoreline of Debidue Island, South Carolina, would not cause downdrift erosion. This was a stunning conclusion about a stretch of coast with a well-documented, consistent north-to-south longshore sand transport. The impossible conclusion was motivated by the need to find the "truth" according to the needs of several very prominent and wealthy community leaders whose houses were about to fall into the ocean. Sensing the disingenuous nature of the engineering report, the community voted down funding for the groins.

Most engineering failures are caused by storms. If they are not, storms are blamed anyway. We hear a common refrain: the seawall or groin or breakwater failed or the replenished beach disappeared because of a large and quite "unexpected" storm that came from an "unusual" direction. Imagine the public response if an engineer who designed an elevated water tank or a skyscraper explained that its collapse was due to high winds! Underlying the public acceptance of the frequent inaccuracy of coastal engineers' predictions is the fact that such engineering is usually paid for by the community at large and not by beachfront-property owners, who created the problem by building too close to the beach.

If one compares pre-project predictions with post-project outcomes of beach engineering, failure would seem to be much more common than success. If other civil engineers had the same rates of failure, it would be an adventure to stand under an elevated water tank or drive across a bridge or stand anywhere near a skyscraper in a windstorm. This being the case, why do coastal engineers have any credibility at all? Why do people continue to believe them and rely on them to hold back the sea? The two big answers are storms and politics.

WHAT ENGINEERS DO

Coastal engineers are called upon for many tasks at the beach. Most commonly, engineers are hailed to save beachfront buildings, to halt a retreating shoreline, to reduce the frequency of flooding, to restore a beach, or to maintain navigation channels. For the majority of these projects, engineers also assess the environmental impacts and the cost-benefit ratios.

Efforts to halt shoreline retreat involve either soft or hard stabilization. *Soft stabilization* involves putting in an artificial beach (beach replenishment) or bulldozing sand from the lower to the upper beach or from one end of a beach to another (beach-sediment recycling). *Hard stabilization* involves the construction of a fixed, immovable structure of some kind, such as a seawall, groin, or offshore breakwater.

The negative environmental impacts associated with hard engineering structures were long ignored, denied, or downplayed, but within the last two decades they have been globally accepted and recognized by engineers. However, these impacts have long been known and understood by coastal scientists. Nathaniel Shaler, a Harvard geologist, recognized as early as 1895 that seawalls destroy beaches, in this case by cutting off the sand-supply source. He also provided the earliest mention of beach replenishment in the United States (albeit with boulders) that we are aware of:

> On the parts of the shore where the land has been extensively occupied by summer residents, the owners have in many cases protected the coast from erosion by embankments and sea-walls, thus diminishing the amount of debris which was formerly contributed to the pocket beaches. In these artificial conditions the beaches often wear out, and the sea begins to assail the part of the coast, which was once well protected. In such cases the only way in which the erosion can effectively be corrected is by carting each year to the beach a sufficient quantity of large bowlders to give employment to the waves and prevent their encroachment upon the shore. The larger these bowlders, the better; for if they are of small size they will be tossed about by slight storms and rapidly wear out, while masses weighing half a ton will be stirred only by the more tumultuous seas.

The British geographer E. M. Ward, in his 1922 book *English Coastal Evolution*, noted both the beach-destroying impact of seawalls and their offshore steepening effect that led to higher waves at the shoreline: "Vertical seawalls were found to induce a scouring away of the beach below them and many a fine beach, like the sands that now cover the once somewhat bare foreshore at Tor Abbey near Torquay, have been lost owing to the erection of seawalls, which later had as a consequence to resist an increasingly violent wave attack. Unsuitable coast defence works on the Dymchurch shore of Romney Marsh led in the early part of the reign of Victoria to a great depletion of the local beach and a violent wave attack."

TREATING BEACHES WITH CONTEMPT

The list of calamitous things that hard engineering structures do to beaches seems endless. Underneath it all is the attitude that the trappings of humans are more important than these beautiful and finely tuned natural features constructed by the waves and the wind at the border between sea and land.

- *Erosion of the adjacent beaches*: Virtually every type of coastal-engineering structure intended to hold the shoreline in place will reduce sand flow from other beaches.

- *Erosion of fronting beaches*: Any structure in the broad seawall, breakwater, and groin family that is placed on the upper part of an already-eroding beach will cause the beach to narrow further and eventually disappear.
- *Increase in the height of the surf-zone waves*: Holding an eroding shoreline in place steepens the nearshore zone, out to a depth of perhaps 20 to 30 feet (6 to 9 meters). This reduces the friction of waves on the seafloor and causes the breaking waves to increase in height, which then increases the rate of shoreline erosion.
- *Reduction of water quality*: Offshore structures often create still-water areas next to the beach. The lack of circulation and exchange can allow pollutants and floating debris to accumulate.
- *Detrimental to turtle and bird nesting*: Structures affect beach fauna in various ways, including removing the nesting beach above the high-tide line and preventing turtles from getting to the beach.
- *Loss of the entire beach ecosystem*: The loss of the beach along with changing wave conditions and sediment type can destroy the ecosystem. And the loss of vegetative wrack in front of walls takes away an important source of nutrients for organisms in nearshore waters.
- *Impede human activities*: Engineering structures can prevent access off of and onto the beach and provide an obstacle to walking or jogging. Storm rubble endangers swimmers and when the beach is lost, recreational use is halted completely; all that is left is the sea breeze.

THE SCOURGE OF BEACHES
Seawalls

Seawalls are shore-parallel walls usually constructed on the upper beach. These hard structures may be constructed of just about anything, but most commonly they are made of steel, riprap (boulders), concrete, gabions (wire baskets filled with stones), sandbags, or timber. In some remote villages in the developing world and a few in the developed world, seawalls are constructed using household trash, such as old refrigerators, dog sleds, derelict cars and trucks, empty oil drums, and even kitchen sinks. These seawalls are designed to reduce erosion during storms, and to prevent or reduce storm overwash. The fundamental raison d'être for most walls is protection of beachfront property.

As a general rule, construction of a seawall is the best way to combat long-term erosion but it is the worst thing one can do to a beach. Virtually every eroding beach with a seawall has already lost or is gradually losing its beach. Of course, few seawalls are built on beaches that are not eroding.

Seawalls create problems in a number of ways. *Passive erosion* happens because walls don't address the cause of the erosion. Thus, after the seawall is built, beach retreat continues—the beach is squeezed up against the seawall and eventually disappears. A 2012 study by geologist Chester Jackson and colleagues on beach width in front of 48 seawalls around the coast of Puerto Rico found that the natural beaches were two to four times wider than the beaches with sea walls. Some of the beaches in front of the walls were completely gone.

Active erosion refers to loss of beach sand due to the direct interaction of the wall and the beach. This occurs because at high tide, seawalls can intensify longshore currents immediately adjacent to the wall, and also because, under the right circumstances, waves are reflected back onto the beach. As the beach narrows, the amount of longshore sand transport is also lessened, reducing the sand supply and causing increased erosion rates on adjacent beaches. Over a time span of several decades or more, seawalls cause the steepening of the deeper parts of the beach (the shoreface), which means that bigger waves with more energy reach the remaining beach. According to geologist Edward Anthony, the beach in front of France's Stes-Maries-de-la-Mer seawall steepened from 0.4 percent to 1.2 percent between 1895 and 2005. Consequently, wave heights in future large storms in the area are expected to increase by 20 percent.

Placement loss happens when seawalls are constructed on the beach, seaward of the high-tide line, as happened on Miami Beach in the 1970s. Before the city made it illegal, hotels competed to have the most seaward-extending seawall and bragged about it in their advertisements. Thus from the day the walls were completed, significant beach loss occurred. Another example is the Victorian-era seaside resort of Aberystwyth in Wales that was repeatedly damaged by storms during the winter of 2013–14. The town's promenade was covered in beach gravel and the road was ripped up by waves. Although there was widespread shock at the damage, a little historical research quickly revealed that the promenade and road had been built on top of the beach. Little wonder, then, that beach pebbles

FIGURE 18 An unusual seawall using concrete tetrapods on Sylt Island, one of the German Frisian Islands. The beach in front of the seawall is replenished and can be expected to disappear eventually. Seawalls, besides causing the eventual loss of the beach, restrict movement to and from the beach. PHOTOGRAPH BY ANDREW COOPER.

were thrown onto it during a storm. It wasn't the first time either—in the archives were photographs of exactly the same type of damage done by previous storms.

Many of the resorts on the Costa del Sol in Spain were built on the wide back-beach during a period of beach stability. When high-energy storm conditions returned, the store of sand in the back-beach had been rendered inaccessible as a sand source for the beach as it was covered with concrete.

Sand supply cutoff on rocky bluffed coasts will cause beach loss if a seawall is protecting a bluff that normally would have provided fresh sand or gravel to the beach as it eroded. Shaler recognized this problem as early

FIGURE 19 Another unusual seawall protecting Kingsgate Castle atop the chalk cliff at Kingsgate Bay, Broadstairs, on the Isle of Thanet in England. The columns, constructed to support the plinth above, are part of the seawall that prevents bluff erosion and stabilizes the cliff top. The columns may also represent an attempt to improve the aesthetics of the engineering structure. The dog in the foreground provides a sense of scale. PHOTOGRAPH USED BY PERMISSION OF GINA LONGO.

as 1895. This situation is a continuing crisis along the coasts of California and New England and everywhere that beach sediment is derived from erosion of soft cliffs.

The most seawalled country in the world, other than Belgium, where almost the entire 41-mile-long shoreline (67 kilometers) has walls, is probably Japan. About 40 percent of Japan's shoreline is lined with concrete breakwaters, seawalls, or rock revetments. The nation gives high priority to preserving land and low priority to preserving beaches, and as a result most of Japan's shorelines with hard structures are beachless or have degraded beaches. Japan has a tradition of building massive walls, and unfortunately Japanese engineers have been instrumental in training coastal engineers throughout Asia, including Taiwan and Korea.

According to a 2013 BBC report, about 44 percent of the coast of England and Wales has some form of engineering on it, but this is the case for only 6 percent of Scotland's coast. The coastal engineering in the United Kingdom, however, is much less "grand" than that used in Japan and some of it is soft engineering—including managed realignment.

Seawalls became front and center in the international news after their failure to halt tsunami waves from Japan's earthquake on March 11, 2011. The largest seawall in the world, at the entrance to Kamaishi Bay, was completed in 2009 at a cost of US$1.5 billion. The 1.2-mile-long (1.9-kilometer) structure (extending into water depths as high as 220 feet or 67 meters) was readily overtopped by 14-foot (4.3-meter) tsunami waves. An even bigger disaster may have been the failure of the wall around the Fukushima Daiichi nuclear-power plant, where a serious and dangerous pollution problem continues to this day.

There are several other famous seawalls around the world. The Cape May, New Jersey, seawall, with a promenade path, is on the first major tourist beach in the United States. Before the seawall was built, there was a healthy beach there, and in 1912 Henry Ford showed off his Model T cars by racing up and down the beach. Since the wall was constructed, even several replenishment projects have not succeeded in keeping a good beach in place, so promenading on top of the wall remains a favorite pastime. The 27-foot-high (8.3-meter) Pondicherry wall in South India was first started in the early 1700s. It gained fame by successfully protecting the community from the 2004 Indian Ocean tsunami. Most seawalls didn't. The beach in front of the Pondicherry seawall disappeared long ago. The 17-foot-high seawall (5.2 meters) in Galveston, Texas, was constructed immediately after the 1900 hurricane that killed 6,000 people. There remains only a small beach, maintained by beach replenishment, in front of the wall. The wall on Ventnor, Isle of Wight, United Kingdom, is a massive one, with a promenade on top and no beach. The immense stepped wall on the seaward margin of the Norfolk Broads in England has degraded much of the beach but is protecting almost no structure of consequence behind the wall. The Maldives island of Male has a seawall (12 feet or 3.7 meters high) that was constructed by the Japanese at a cost of US$60 million. Most of the former beaches were lost, but some have been

FIGURE 20 The Galveston, Texas, seawall was completed in 1910 as a response to the 1900 hurricane that killed at least 6,000 people. At the same time that the seawall was constructed, the elevation of much of the community was raised by 17 feet (5 meters) with the addition of new sand. First floors of business buildings became basements. PHOTOGRAPH BY ANDREW COOPER.

replaced by beach replenishment. The wall has caused the city to lose some of its attraction for tourists. The Sunabe seawall on Okinawa, Japan, is a large concrete structure — with no beach. The O'Shaughnessy seawall at Ocean Beach, San Francisco, was built in 1928. It is a large, stepped, concrete structure similar to the Galveston wall, but it is a rare example of a major seawall that has not caused beach loss. This is likely due to a large and continuing natural sand supply. In Porthcawl, Wales, the beach is backed by a two-tier seawall. While it protected the road, it contributed to narrowing of the beach. The community decided to combat this by covering much of the beach with tarmac, an example of cutting off the nose to spite the face. Imagine sunning yourself at the beach on tarmac! Sadly, it isn't the only instance of covering a beach with foreign material. For example, engineers have done the same thing on the German Frisian

Islands. On Wangerooge Island parts of the beach have been covered with a gently sloping concrete slab to halt erosion.

Groins

Groins, usually made of concrete, wood, stones, or steel, are as widely used as seawalls. Built perpendicular to the shoreline, these structures trap sand flowing along the beach and widen the beach on the updrift side of the structure, but they reduce the sand flow to adjacent downdrift beaches. Groins can also be made of many other kinds of materials, ranging from a line of dump trucks and cars (e.g., in Majuro, Marshall Islands) to a wall of massive boulders (e.g., in Espinho, Portugal). There are many types of groins, including high groins, low groins, T-shaped groins, biodegradable groins, fish-net groins, permeable groins, and temporary removable groins.

Because of the erosion caused by groins, adjacent property owners are often forced to install groins in self-defense, all of which eventually leads to large numbers of groins (a *groin field*) on a shoreline reach. Groins are sometimes referred to as jetties (e.g., in New Jersey), and the term *terminal groin* has been used to describe the last groin in a groin field at the downdrift end. At present the term *terminal groin* is incorrectly used to describe some jetties at navigation inlets.

Groins create other problems for beaches besides downdrift erosion. Like seawalls, groins are ugly, at least relative to an unengineered natural beach. And groins, like any coastal-engineering structure, need maintenance, an often-ignored fact. Lack of maintenance leads to a beach littered with fragments of wood, stone, or concrete scattered about that sometimes endanger swimmers. Groins present an obstacle to walking or jogging on beaches, and people require skills in rock climbing to move from one groin compartment to another.

Geographer Keith Clayton, commenting on engineering works at the former sandy beach of Happisburgh in Norfolk, England, notes: "The beach is now broken up into compartments by ugly, high groynes and a revetment. At high tide a beach user trapped behind the revetment cannot see the sea. At mid-tide each compartment can be reached by a somewhat perilous clamber over the revetment. . . . Once on the lower beach the view in three directions is of a wall of timber. . . . The decayed gabions and lengths of railway line below the cliff, together with the revetment[,] have done little more than reduce the rate of cliff retreat."

One of the most severe beach-erosion problems in the United States was caused by groin construction at Westhampton, New York, on Long Island. In the middle of the project, the funding for a planned groin field disappeared, so the half-finished project was left to the elements. The groin field was constructed from the updrift end first, a major blunder, so when the project was halted, there was a long stretch of sand-starved beach downdrift. It was bad engineering caused by bad politics. Would civil engineers allow cars to cross an unfinished bridge? Or build a dam across an active fault? Or construct a faulty building because money ran out? The elements (in this case, storms) quickly removed some houses and formed a new inlet beyond the last groin. The total cost of this engineering error helped along by local politicians will soon exceed $100 million.

The west coast of Portugal is subjected to high waves that push large volumes of sand mostly toward the south. As a result, when sand supplies are interrupted by groins, shoreline erosion to the south is immediate and rapid. According to geologist Helena Granja, most of the critical erosion problems along the west coast of Portugal are a direct result of coastal-engineering structures. This is coastal engineering at its worst. For example, the groin created in 1986 in Pedrinhas, a mile south of a hotel, did not build up the hotel's beach as predicted, but it did cause immediate and severe erosion downdrift on a shoreline that had previously been stable. This erosion threatened beachfront houses, so a boulder revetment was constructed along the shoreline. Over the years the groin has been repeatedly elongated as the beach retreated, and the beach in front of the houses has long since disappeared—all this on a coast that was not eroding originally or was eroding very slowly before the arrival of the engineers.

Interestingly, groin compartments often provide the basis for separation of beach goers. Child-friendly beaches may be segregated from surfing beaches, gay beaches, nude beaches, volleyball beaches, and so on. This was the case for a beach in Barcelona where a public hue and cry occurred when a groin field was removed and the beach was replenished.

One group of beachgoers may consider groins a benefit. Experienced surfers often ride the strong seaward-directed currents that can occur next to groins to get out to the big waves. Unfortunately, these same currents can carry inexpert or unwary swimmers out to deeper waters, putting them in danger. During big storms, currents parallel to groins are strengthened by the weather conditions and tend to carry large amounts of sand out to sea, adding to the beach-erosion problem. In Durban, South Africa, some of the groins have been so undermined by the scouring of offshore-moving currents that they have been closed to public access, pending replacement or repair.

Jetties

Jetties are walls built perpendicular to the shoreline adjacent to inlets or entrances to harbors. Their purposes are to prevent sand from clogging navigation channels, to reduce channel-dredging costs, and to make navigation a bit safer. Jetties have caused some of the largest shoreline-erosion problems in the world. For example, the jetties at the south end of Ocean City, Maryland, have cut off the sand supply to Assateague Island to the south. The erosion is so great that the northern half of Assateague has migrated its entire width. The front side of the barrier island is now landward of what was the backside or bayside of the island in 1933, when a hurricane opened the inlet.

Some short jetties are now being referred to as *terminal groins* and are purported to help alleviate the erosion problem of sandy shorelines. The change in terminology was basically an attempt to get away from the bad erosion reputation that has come to be associated with the term *jetty*. The ridiculous assertion of the engineering community in North Carolina that terminal groins will not cause downdrift coastal erosion has caused the unraveling of the state's long-standing anti-hard-stabilization regulations and has far-reaching implications for the state's beaches.

Breakwaters

Offshore breakwaters are seawalls built in the water parallel to the shoreline. They can reduce the height of waves during storms and cause sand to be deposited behind the breakwaters. As with groins, the beach widens immediately behind breakwaters, where the beach is sheltered from waves, but adjacent beaches erode because of loss of sand. If the widened beach behind the breakwater extends like a horn all the way out from the beach to the structure, that sand body is called a tombolo. Breakwaters off the beach in Dunkirk, France, where the tide range is as high as 17 feet (5.2 meters), cause mud to be deposited behind them. In fact, breakwaters are not commonly used on France's Atlantic coast because of high tidal ranges. However, offshore breakwaters are a common coastal-engineering structure in Japan and have been widely used in other parts of the world (e.g., Presque Isle, Pennsylvania; and the coast of Sicily).

Engineers claim that if the spacing between the breakwaters is correct, depending on local wave conditions, sand will continue to flow behind the structures. This very rarely happens, because along any given beach a wide variety of wave conditions will occur. In other words, the breakwater spacing designed by engineers may be fine for waves from one direction but not others. This was found to be the case in Holly Beach, Louisiana, known in some circles as the Cajun Riviera, where breakwaters that were constructed to halt shoreline retreat failed to do so. An engineering firm calculated that the spacing was incorrect, so at great cost the old breakwaters were removed and new ones were put in place. Along came Hurricane Rita a few years later, in 2005, and the small town was virtually destroyed. Only the breakwaters remained!

Giorgio Anfuso from the University of Cadiz cited the 56-mile (90-kilometer) shoreline of the Ragusa Province of southern Sicily as an example of bad coastal management. A sequence of events occurred "where one unwise action was countered with another, which in turn created additional problems. The situation arose through strong political interference and ignorance (or lack of concern) regarding the environmental impacts of human interventions on the shoreline and by the public perception that government has a duty to protect private property." The coastal retreat caused downdrift of the first breakwaters was countered by the emplacement of yet more breakwaters creating a "domino" effect. To make matters worse, the breakwaters were constructed "to protect unplanned and illegal . . . beachfront summer houses. Without the presence of these houses there would have been no need for publicly funded intervention."

One of the largest breakwater projects in the United States is on Presque Isle, Pennsylvania, on Lake Erie. Presque Isle is a 7-mile-long (11-kilometers) sand spit that extends into the lake and is a heavily used state beach. The spit had a minor erosion problem that was being handled by beach replenishment, but the U.S. Army Corps of Engineers decided to build 55 offshore breakwaters along the spit to hold the shoreline in place. The engineers stated that the structures also would allow sand to flow to the unarmored tip of the spit, known as Gull Point, an important and relatively pristine nature preserve. The promotion by the Corps of this nationally visible and highly controversial project became a decades-long juggernaut.

Scientists and engineers at the Coastal Engineering Research Center in Biloxi, Mississippi, the coastal research arm of the Corps (the ones who should know better), expressed certainty in the design assumptions and agreed that sand would flow behind the breakwaters. This impossible conclusion was reached through the use of a mathematical model. In 1978, during the design phase, the Corps built three prototype breakwaters as an experiment, to be sure that sand would flow behind the structures. Sand didn't, and the breakwaters caused immediate severe downdrift erosion. The Corps engineers attributed the erosion to a storm rather than to the breakwaters. To top it off, before the first stone splashed into Lake Erie, an outside engineering consulting firm hired by the state of Pennsylvania had warned that the project would not work as intended.

In 1993 the National Society of Professional Engineers awarded the U.S. Army Corps of Engineers its Outstanding Engineering Achievement award for the project. In 2011 the American Shore and Beach Preservation Association declared the beach to be a huge success, and the Association's 2013 newsletter lists Presque Isle as the best park or habitat beach in the nation.

In actuality, today the beach is suitable for swimming, provided one doesn't stray too close to the boulder breakwaters emplaced in 1992. But continued beach replenishment at unpredicted rates and volumes has been required to maintain the beach. Sand never did flow behind the breakwaters in the way that was anticipated, and frequent beach bulldozing is required to remove the tombolos. And, the all-important Gull Point has been severely eroded and will likely disappear altogether in the future. The open lake-facing shoreline on the point had retreated 500 feet (152 meters) by 2009, and tangled trees and shrubs covered the beach and nearshore zone.

The project proved to be an environmental disaster just as predicted by most outside observers, but the temptation to engineer the shoreline with one of the largest projects of its kind proved too strong. The beach, Gull Point, and the truth be damned.

FIGURE 21 Sandbag seawall on North Topsail Beach, North Carolina. Sandbags were allowed by the state when it outlawed hard stabilization, but the impact of sandbags on the beach is no different than that of a vertical concrete wall—that is, the beach will eventually disappear. Sandbags, however, are relatively fragile and are often destroyed in storms. PHOTOGRAPH USED BY PERMISSION OF DUNCAN HERON.

Geotubes

Geotubes are large cylindrical sandbags that are used for many coastal-engineering purposes. The bags, which are typically several feet (a meter or two) in diameter and tens of yards (tens of meters) long, are made of geotextiles, cloths through which water will flow without loss of sand. The sand filling the tubes is usually local beach sand. Geotubes are often substituted for stone, steel, and wood engineering structures and have the advantage of being relatively easy to remove. The bags are fairly fragile, however, since they are susceptible to ultraviolet radiation, vandalism, storms, birds that treasure fabric thread, and theft by people who use the

FIGURE 22 Galveston, Texas, geotubes did not survive well in Hurricane Ike in 2008, although they probably reduced the amount of overwash and erosion during the storm. The beach in front of the geotubes was narrowed after the storm. PHOTOGRAPH BY ORRIN PILKEY.

cloth for sunshades. When abandoned but not removed, the bags can be a significant hazard to swimmers. Geotubes have been used as seawalls (e.g., in Galveston Island, Texas), groins (e.g., in Bald Head Island, North Carolina), offshore breakwaters or reefs (e.g., in Queensland, Australia), dikes (e.g., in Mobile Bay, Alabama), and submerged breakwaters (e.g., in Marseille, France). A very common application in Florida and Portugal is so-called dune restoration, where a geotube on the uppermost beach is covered with sand and then planted with grass.

Other factors being equal, geotubes do not differ in the slightest from concrete, stone, or steel structures in their impact on beaches. The dune-restoration use is particularly absurd since the sand covering is often removed permanently by wind or waves and the bags are exposed after the first significant storm. More and more of the world's beaches are littered with tattered remnants of bags that are difficult to remove because they are embedded in beach sand. Figure Eight Island in North Carolina has remnants of three generations of geotubes visible, all on the same section of beach. Each generation of geotubes had been destroyed by storms.

Geotubes are not soft stabilization, although this is a common claim. Their principal advantage is that people can remove them, but if they are not removed, they are soon destroyed by nature, often within two or three years. In the meantime they remain an eyesore and cause problems just like other hard structures.

Gabions

Gabions consist of wire cages or baskets filled with rocks that are stacked on top of one another. They are a low-cost and easy-to-install approach to beach preservation, best suited for coasts with low wave energy. They are a common type of seawall, particularly in island nations and communities where large rocks are not available. For example, gabions are frequently used for stabilization on atolls, at Inupiat coastal communities in Alaska, and in Puerto Rico. Because of saltwater corrosion of the wire, gabions have a short life span (three to six years). When they begin to disintegrate, they not only degrade the beach but also sometimes form very serious hazards to persons on foot as the enclosed stones spill onto the beach accompanied by wire fragments.

Snake-Oil Devices

In the world of coastal engineering is a whole army of devices backed up by an army of engineers, each with a story of how a particular device will solve the erosion problem without damaging adjacent beaches. The names are as ingenious as the devices. They include Surge Breakers, Sea Scape, Reef Balls, Wave Wedges, Pep Reefs, Biorock, Beach Saver Reef, Beach Cones, Beach Prisms, and Wave Shields. Some are named for their inventors, such as the Holmberg Undercurrent Stabilizer, the Parker Sand Web, Stabler Disks, and the Menger Submerged Reef.

Snake-oil devices basically perform the same functions as other hard structures. They attempt to block the waves like a seawall, or trap sand like groins and breakwaters. These devices also cause the same damage to adjacent and fronting beaches as traditional structures do. Snake-oil devices frequently come with fantastic claims of how they work and of past successes, often in faraway lands. The Holmberg Undercurrent Stabilizer

FIGURE 23 Septic-tank rings in front of the Royal Atlantic Hotel in Montauk, New York, being installed without a permit. These were chained together, intended as a temporary storm and erosion barrier, but they were easily broken up in the first nor'easter. PHOTOGRAPH USED BY PERMISSION OF JOHN H. CHIMPLES.

(which is basically a groin) claims to bring sand ashore to the beach and was said to be very successful in Saudi Arabia.

These devices are generally on a smaller scale than the more standard structures, can be more easily removed, and often clutter beaches, making swimming more dangerous. For example, the Parker Sand Web consists of netting extended across the beach and attached to short poles. A real danger exists, of course, of someone getting tangled up with the nets. Other devices are described as permeable, adjustable, and even biodegradable. A few are designed for emplacement immediately before storms, then to be removed after the storm passes. Many, if not most, include an element of sand trapping often accomplished by reducing the size or the velocity of the backwash in the surf zone. One patented device is supposed to reduce backwash by removing water from the swash, and another proposes to reduce longshore transport by pumping water against the current.

Claims of success must always be evaluated carefully. One snake-oil device, Seascape, consists of artificial kelp-like seaweed or strips of plastic extending into the water column and anchored by a bag of sand to the seafloor. The idea was to baffle or reduce the wave impact and cause sand carried by incoming waves to be deposited to build up the beach. Shortly after the bags were placed in the nearshore zone in front of the Cape Hatteras Lighthouse in North Carolina, the beach suddenly widened. This is true enough. But the cause of the beach widening, which simultaneously occurred along 50 miles (80.5 kilometers) of shoreline, was a few days of relatively strong winds from the southeast, an event that always causes temporary beach widening on the Outer Banks. The declaration of success by the Seascape promoters was premature.

The claims of those selling such devices sometimes overwhelm common sense. In Recife, Brazil, a contract was recently signed to emplace a line of "Seabags." These sandbags covered with geofabric were supposed to attract sand and promote natural beach recovery. Those grandiose and obviously untrue claims were made in spite of the fact that a previous 4,000-foot-long (1,200-meter) installation of the same sandbags on an adjacent coast collapsed shortly before the new contract was signed.

Desperation Beaches

The public is often outraged when a beach has been lost in front of a seawall. To placate them, we have seen many instances where a narrow strip of sand is placed on top of or landward of the seawall, sand that will never see a wave, except in a storm. Boa Viagem beach in Recife has been destroyed and a 10- to 15-foot-wide (3- to 5-meter) strip of sand has been placed next to and landward of the seawall. Squeezed into this space are beach volleyball pitches with nets to prevent the players from falling off the wall. Similar faux beaches include Seabrook, South Carolina; Pondicherry, India; and Nusa Dua, Bali. Although they are next to the sea, these beaches lack the cleansing of seawater and can quickly become polluted as cats and dogs use them as toilets.

Beach Replenishment

Beach replenishment involves placement of sand from an outside source on a beach. It is costly (a million dollars per mile at a minimum in the United States) and temporary (typically of two to seven years' duration, depending on location), but it has the advantage of supposedly improving the beach rather than destroying it. Replenished beaches are discussed in the next chapter.

A BANKRUPT PROFESSION?

We have seen a small sampling of instances where coastal engineering has damaged or destroyed beaches. It is true that in most cases coastal engineers on beaches are acting in the interest of their customers to save buildings, roads, and other community infrastructure, a noble pursuit by any measure, but there is a murkier side to the profession. This stems from several linked issues that broadly follow the sequence of events involved in implementing an engineered intervention on the coast from conception to ultimate failure. The system, it turns out, is rotten from start to finish for the following reasons.

Coastal behavior cannot be predicted with the accuracy claimed. Beach engineers claim an ability to predict the future of beaches (using numerical mathematical models), an ability that is truly nonexistent. They routinely make predictions through rose-colored glasses about the nature and extent of the impact of structures, and even about how long the nourished beach will last. Numerical models are gross simplifications of reality, and they cannot predict beach behavior. As we saw in chapter 1, beaches are affected at any given time by a whole variety of processes (e.g., individual waves, tidal cycles, storms, seasonal variations in climate, and sea-level change) that interact with each other in a chaotic way to produce the net changes on the beach. None of these individual processes (let alone their interactions) can be represented by an equation—yet model simulations are routinely presented as accurate and state-of-the-art predictions.

Predictions of shoreline behavior can be manipulated to suit the requirements of the project. In light of the fact that beach behavior cannot be accurately predicted, it is straightforward to manipulate models to make a prediction according to the desired outcome. Plans to nourish a beach

FIGURES 24, 25, 26
Three examples of
beaches artificially
constructed on top
of seawalls that have
destroyed the natural
beach. These are a very
poor substitute for a
genuine beach. They are
Pondicherry, India; Recife,
Brazil; and Nusa Dua, Bali.
This may be the wave of
the future, as more and
more beaches in tourist
communities will have
seawalls. PHOTOGRAPHS BY
ANDREW COOPER; NUSA DUA
PHOTOGRAPH FROM ORRIN H.
PILKEY, ET AL., *THE WORLD'S
BEACHES*.

at Nags Head, North Carolina, ran into problems over concerns that sand would be transported downdrift and fill in the nearby inlet (Oregon Inlet). In mathematical-model studies during the design phase, the original assumed angle of wave approach was reduced by 5 degrees, which immediately minimized the longshore sediment-transport rates and solved the problem—at least from the point of view of ensuring that the project went ahead. In reality, of course, more than a decade later the inlet *is* filling in and has to be dredged frequently. Whether one wishes to consider it irresponsibility, dishonesty, or incompetence, the willingness to bend the facts to support untenable or damaging projects seems to permeate the coastal-engineering profession. This includes private engineering and science consultants who take advantage of a booming consulting industry.

An engineer is most likely to offer an engineered solution to any problem. Usually when engineers are asked to report on a coastal-erosion issue, one or more "solutions" are offered. Inevitably it is in an engineering company's interest to propose an engineered approach since that means more business. If the potential budget is not known, several options are usually offered that span a range of prices. In the event that the budget is known, engineered solutions are designed to take maximum advantage of the available budget.

The size of the engineering intervention depends on the size of the budget. There are many examples of over-engineered beaches where the local authority had more money than sense. It seems possible to expand any project to fit the available budget.

Failure is blamed on acts of God. The inability to predict beach behavior is often only evident when projects fail. In these circumstances, "unusual storms" are the usual scapegoat. This is particularly evident in beach-replenishment projects when beaches fail to survive for as long as originally predicted.

Failure means more business. Unlike in most other professions, failures in coastal engineering usually lead to more business. This arises from the ability to blame unexpected events or the client having selected a cheaper option. More business is generated with more engineering works in order to undo past mistakes (e.g., restore failed structures or replenish beaches again) or deal with the consequences of structures (e.g., extend sea defenses downdrift to cope with erosion caused by the original structures).

There is no looking back. The routine lack of monitoring of coastal-engineering projects has two serious consequences. First, no lessons are learned from past mistakes. Second, nobody is held accountable for failures. All these things lead to considerable damage to whomever foots the bill (commonly taxpayers) and to beaches themselves.

THERE ARE CLEARLY SERIOUS FAILINGS in the way in which coastal engineers interact with beaches, the public, and those with responsibility for beach management. No two countries have the same approach, but, despite differences in detail, the well-documented failures of the profession in the United States are symptomatic of coastal engineering everywhere. The fundamental problem with U.S. coastal engineering is the federal agency that must approve all beach projects and usually carries them out—the U.S. Army Corps of Engineers. The Corps is a controversial agency, much loved by development interests and much distrusted by environmentalists.

It is an agency that receives funding on a project-by-project basis. No local projects, no local branches (districts). In that regard it is no different from any coastal-engineering company—but it is quite different from most government agencies. This is an unfortunate arrangement that makes the Corps a project-hungry agency in the extreme, a situation that does not favor honest and competent engineering.

The failed levees in New Orleans during Hurricane Katrina in 2005 are only one example of the agency's incompetence. The levees turned out to be at different elevations at different places. The levees were breached at 53 locations, and, once breached, the unarmored levees were eroded by the cascading water. Ironically, much of the flooding of New Orleans was caused by a raised water level (storm surge) in the Mississippi River–Gulf Outlet (MRGO) navigation channel, a costly Corps of Engineers boondoggle that was built for but rarely used by ships. The channel extended to the city's levee, and the storm surge roaring in from the Gulf of Mexico topped the levee in spectacular fashion.

Driven by its eternal need for projects, the Corps basically cannot be trusted to provide an unbiased view. Instead it has become an arm of the U.S. Congress. The Corps is viewed as a trusted agency anxious to please those with the purse strings and willing to provide suitable, but sometimes questionable environmental-impact statements and cost justifications for projects. Some people argue, with apparent justification, that members of Congress, in their constant efforts to bring home projects for their districts, are primarily responsible for the engineering incompetence of the Corps.

Of course, coastal engineering is not the only profession with a shaky reputation. There are a number of levels of disingenuity that are at least implicitly accepted and even expected by contemporary society. Politicians and lawyers are the first to come to mind, possibly along with used-car and real-estate salespeople.

At the very least, one expects a statement from a politician to have a spin related to his or her political bent. Law students are taught never to lie in court proceedings, but it would appear to be a lesson that has not always been followed. Putting a spin on a courtroom issue is considered legitimate, but the distinction between spinning and lying can be quite fuzzy.

Courtroom manipulation is one thing and political chicanery another, but what makes the dishonesty of coastal engineers unique is that in most of their efforts on the world's shorelines, they are providing answers that cannot be given even by the most skilled and honest of engineers and scientists. That is, there is no absolute truth that a human being can discern about the future behavior of a beach. The engineers are claiming unknown factors as accuracies, which then provide the basis for expenditures of millions of dollars (usually from taxpayers) and emplacement of structures that destroy beaches.

As engineers are constantly being reminded from their efforts at the shoreline:

God always forgives
Man sometimes forgives
Nature never forgives.
(ATTRIBUTED TO ST. AUGUSTINE)

4 PATCH-UP JOBS

Beach Replenishment

BEACH REPLENISHMENT involves placing sand on eroding beaches to restore what has been lost. It originated because of disappearing beaches in front of seawalls, although the connection between walls and erosion was not always recognized. In the United States, the now-giant national replenishment program began as a result of beach loss in New Jersey after the Ash Wednesday Storm of 1962. By 1965, replenishment was widespread in New Jersey, and soon the method was widely used on beaches from Alaska to Puerto Rico and from New York to Hawaii.

In Sea Bright, New Jersey, the shoreline was held in place for more than 100 years by seawalls and groins. The structures were built larger and larger by the decade, and as they suffered wave damage they were replaced. Soon the useless beach became the repository of fragments and pieces of broken walls and groins. In 2002, a 21-mile-long (34-kilometer) replenished beach was installed, extending south from Sandy

Hook beyond Sea Bright to Asbury Park, at a cost of US$13 million per mile (1.6 kilometers). This replenished beach covered up the narrow, trashed beach.

Replenishment is costly—too costly to be a global long-term solution. According to Andy Coburn of the Program for the Study of Developed Shorelines at Western Carolina University, the initial cost of replenishing the 120-mile (193-kilometer) developed shoreline of New Jersey would be $400 million. Maintaining the state's beaches for 50 years at a cost per mile (or 1.6 kilometers) per year of $866,000 will come to $5.2 billion. Multiply this by the 3,000-mile-long U.S. barrier-island shoreline (about 4,830 kilometers) and the number becomes unfathomable.

Recognition of the damage inflicted on beaches by seawalls means that it is now common practice to proceed directly to beach replenishment on eroding shorelines and to carry out repeated replenishment of beaches. According to Coburn, Wrightsville Beach, North Carolina, has had 19 replenishments and Carolina Beach has had 28 since 1965. Replenishment has become the shoreline-stabilization method of choice throughout the United States and much of Europe. On the German barrier island of Sylt in the North Sea, replenishment was adopted in 1972 after years of failed hard-stabilization efforts. Between 1972 and 2000, the 25-mile-long shoreline (40 kilometers) was held in place with 18 replenishments. In total, this involved almost 39 million cubic yards (30 million cubic meters) of sand at the staggering cost of almost US$158 billion (€115 billion), i.e., each 3-foot length (one meter) of beach cost US$3.85 million (€2.8 million).

We anticipate that beach replenishment will continue until the cost of holding a shoreline in place becomes prohibitive due to sea-level rise, lack of suitable sand, and lack of government funding because of changing priorities. As major cities become threatened by sea-level rise, it is certain that protecting them will be a higher priority than preserving tourist beaches. At that point, seawalls will be used to protect buildings on tourist beaches—and the price will be the eventual loss of the beach. It may seem strange that adding sand to beaches will eventually lead to the loss of beaches, but we predict that almost all replenished beaches will someday revert to seawalls.

Since its first application, beach replenishment has been seen as a way of undoing past mistakes, and it has come to be viewed by many as a panacea for all beach problems. Some proponents even go as far as claiming that this is in some way working with nature—sadly it isn't. The prevailing view is: if we make a mess of a beach, we can always sort it out by adding fresh sand.

If only it were that simple! In fact, not only is beach replenishment costly and unsustainable but it encourages more dense beachfront development. Where there is money to be made, the support for continuing government funding of replenishment is strong. The situation is exacerbated by the fact that many beachfront-property owners are wealthy and influential people who can influence decisions regarding public funding of beach replenishment. Ultimately, once replenishment starts, communities are locked into a vicious cycle from which there is no exit. As sea levels rise and the cost of holding back the sea with multiple replenishments becomes too great (or the task exceeds the technical capabilities of engineers), replenishments will be abandoned and seawalls will be the last-ditch effort to save property. On retreating shorelines, as we have seen, seawalls destroy the beach.

WHAT IS BEACH REPLENISHMENT?

Beach replenishment involves taking sand or gravel from an outside source and dumping it along a retreating shoreline to widen or replace an eroding beach. In this way, replenishment seeks to address one of the issues that contributes to beach erosion: a shortage of sediment. The first major replenishment in the United States was at Coney Island, New York, in 1922. Beach replenishment is now used all over the world, in a variety of countries, including Brazil, Nigeria, Korea, Japan, Ghana, South Africa, Singapore, Malaysia, Western Europe, and various Caribbean nations.

According to Coburn, since 1970 the United States has replenished beaches 469 times using more than 370 million cubic yards (283 million cubic meters), at a cost in today's money of $3.7 billion. The average cost of sand in the United States (which varies widely from beach to beach) has increased from $1.70 per cubic yard (0.8 cubic meters) in 1970 to $14.38 per cubic yard today. Costs of replenished beaches vary depending on abundance of local high-quality sand, dredge operating costs, distance of sand source from shore, and wave conditions at the dredge site. As a rule, 5-foot waves (1.5 meters) at the dredge site halt all operations. Costs will continue to rise as demand for beach replenishment increases and suitable sand has to be sought farther offshore.

FIGURE 27 A sand-water slurry from the adjacent continental shelf being pumped up on Atlantic Beach, North Carolina. After the sand is pumped onto the beach, bulldozers spread it out. PHOTOGRAPH USED BY PERMISSION OF GREGORY RUDOLPH.

A 2002 study of beach-replenishment practices in Europe indicates that since the first operations in the 1950s, around 458 million cubic yards (350 million cubic meters) have been placed on Western Europe's beaches (Denmark to Italy). Two-thirds of that total amount was emplaced on Dutch and Spanish beaches in almost equal amounts. The Dutch are committed to holding their shoreline in its 1991 position for coastal defense purposes. In Spain, most replenishment is on recreational beaches that are developed to accommodate tourists.

Beach replenishment goes by a number of pseudonyms, including beach nourishment, beach fill, dredge-and-fill, artificial beaches, and, as of 2013, coastal storm-damage-reduction projects. The stated purpose of the use of the last term is to make beach-replenishment funding more palatable to inland dwellers—*beach replenishment* seems to imply helping wealthy people enjoy themselves down at the seaside, whereas *storm-damage reduction* recasts them as innocent victims of disasters.

Sand for replenishment is commonly piped ashore as a slurry of sand and water. Source areas include the adjacent seabed, inlets, tidal deltas, and lagoons, as well as inland sources from which sand is trucked in to

FIGURE 28 Dump-truck replenishment in Dubai. Sand is obtained from the Persian Gulf floor, stored in large piles, and then trucked to individual beaches. PHOTOGRAPH BY ANDREW COOPER.

the beach. A decade ago beach replenishment was carried out in Virginia Beach, Virginia, by trucking sand from nearby sand deposits during the tourist off-season. More than 100,000 dump-truck loads in a single year were placed on the beach (more than a million cubic yards or 765,000 cubic meters), but the process was halted in part because the city's streets were being destroyed by the truck traffic. The same thing happened to the streets of Myrtle Beach, South Carolina, before the city gave up using dump trucks for transporting replenishment sand.

Trucking sand still happens in many locations, including the tourist beach of Cesme in western Turkey, but in some places sophisticated engineering works have been put in place to pump sand from one beach to another. On the tourist beaches of Durban in South Africa and the Gold Coast of Australia, as well as at several inlets in Florida, there are systems of pumps and pipes that take sand from one side of a tidal inlet and pump it out on the other side. These cases of *inlet bypassing* are attempts to address erosion problems caused by jetties.

FIGURE 29 Piles of sand that have been bulldozed up from the beach to protect houses on Topsail Island, North Carolina. Such piles of sand up against the houses are usually swept away in the first storm, and of course the ecosystem of the beach is destroyed by the bulldozer. PHOTOGRAPH USED BY PERMISSION OF DUNCAN HERON.

Frequently, the stated justification for beach-replenishment projects is improvement in the quality of the beach or storm protection, but rare indeed is the replenishment project that is not politically inspired by influential owners of beachfront property that is threatened by erosion, whether houses, condominiums, apartments, hotels, or entire resorts. Buildings don't cause shoreline erosion, but buildings do cause the shoreline-erosion *problem* when they are threatened. This being the case, public funds are often used to bail out affluent beachfront-property owners from a problem they created by being there.

As in all things, or at least so it seems, there is a good, a bad, and an ugly side to beach replenishment.

THE GOOD
Beach Replenishment Slows Down Erosion

A new beach temporarily halts erosion and simultaneously affords protection for beachfront property against moderate storms as the upper part of the beach absorbs the impact of storm waves. *Temporarily* is the key word here, as adding sand doesn't solve the erosion problem, and the process has to be repeated for as long as the property is to be defended.

Beach Replenishment Reduces Storm Surges

The artificial beach may absorb or block the storm surge and prevent flooding from relatively small storms. For large storms, however, most artificial beaches have little if any impact on flooding caused by the storm-raised sea level. Waveland, Mississippi, has an artificial beach, which changed little in appearance with the passage of Hurricane Katrina in 2005. The beach was untouched as the 30-foot (9-meter) storm surge simply stepped over the beach and wiped out almost every building in the community's first six blocks, all single-family homes.

A side note: less than a year after Katrina, some small beachfront lots in Waveland, with only concrete slabs to indicate the position of former houses, were for sale for $700,000 and up. In several cases, the beachfront lots had lost their houses in both the 1969 passage of Hurricane Camille and the 2005 passage of Hurricane Katrina. Today some of these lots sport costly new houses awaiting the arrival of the next storm.

In New Jersey, Hurricane Sandy in 2012 showed that a wide beach and, in particular, a high artificial dune sometimes reduced property damage from the storm surge. Now, on a statewide basis, the New Jersey communities are touting high artificial dunes as the "solution" to future storms. But one big problem has arisen. Before building the dunes, all property owners must agree to allow a dune. A number of homeowners object to the artificial dune because it ruins their sea view. In Harvey Cedars, 18 of 82 residents refused to sign the requisite papers to allow construction of a dune in front of their houses. The project went ahead anyway and Harvey and Phyllis Karan sued the city because the dune blocked their view, even though the sea was still visible from the second and third floors of their house. In an astounding lower-court decision, the couple was awarded $375,000 to compensate for the diminished property value caused by the lost sea view. The city appealed and the award was thrown out. A coastal engineer (not identified in media reports) whose testimony was not allowed before the jury in the appeals court stated that the dune would protect the property for at least 200 years. A more absurd statement about

FIGURE 30 This beachfront property in Mississippi was for sale for US$700,000. The house that was on this lot disappeared in Hurricane Katrina in 2005. The first house on the lot disappeared in Hurricane Camille in 1969. Who would be foolish enough to buy this lot? PHOTOGRAPH BY ORRIN PILKEY.

dune durability is hard to imagine. If a storm strikes the area the day after the dune is constructed, the dune could disappear instantly.

Beach Replenishment Provides Recreational Opportunities

The new beaches provide recreational opportunities that didn't exist on the narrowed beaches, assuming that the quality of the sand is good. (The quality problem is a big *if*, as discussed below.) In many European countries where patches of beach are leased to traders (cafés, deck-chair rentals, and so forth), the bigger the beach, the better it is for the municipality. Similarly, when a beach attracts an admission charge, as is common on the French and Italian Rivieras and in New Jersey, squeezing people onto the beach means that more money can be made. There are limits, however, as some replenished beaches in Spain are so wide that the back beach becomes too hot to be comfortable. It is abandoned and the sunbathers confine themselves to a narrow strip close to the water.

From a surfer's standpoint, nourished beaches can ruin surf breaks. Many instances of this have been documented in New Jersey, and the world-famous surf break at Kirra, Australia, was lost for a period when too much sand was pumped onto the beach, wiping out the wave break.

Beach Replenishment Is Better Than Seawalls

Replenishment has one huge advantage over seawalls, groins, and other forms of hard stabilization that degrade or destroy the beach. Instead, replenishment ensures that a beach continues to exist in one form or another. In some cases, sand from artificial beaches moves laterally and furnishes fresh sand to adjacent beaches. A notable example occurs on the German North Sea island of Sylt, where sand from decades of replenishment has leaked northward, causing the widening of a beach in neighboring Denmark.

Whether or not a beach is improved by replenishment depends on how close the grain size of the new beach is to the old. Having a similar grain size is important to ensure that the ecosystem that may eventually return will be the same as the original. It can hardly be characterized as an improvement if the sand on the new beach becomes cemented, as in some replenished beaches in Spain and in parts of Miami Beach, because of the fine lime mud that is pumped along with the sand, or if a former fine-sand beach changes to one containing a lot of mud or shell gravel (as on Bogue Banks, North Carolina).

THE BAD
Beach Replenishment Is a Band-Aid Solution

Replenished beaches disappear—in most cases much faster than natural beaches. Fundamentally, this is because a replenished beach, even a very large one, is just a small sliver of sand when compared to the entire beach, which can extend far offshore into deeper water. Studies indicate that most replenished beaches disappear at least twice as fast as the natural beaches that preceded them. Thus, when a beach is replenished, it must be viewed as the beginning of a long and never-ending process of repeated replenishments extending into the distant future, until sea-level rise makes it economically impossible to hold the beach in place, or until the sand source is used up. A real Band-Aid on a child's knee may have to be replaced every few hours; a beach must be replaced every few years.

In a very general way the life span of beaches is determined by the average energy or height of the waves. Thus, beaches on the east coast of Florida last longer (an average of seven to nine years) than the beaches of New Jersey or the Outer Banks of North Carolina (an average of three or four years). In addition to wave height, the grain size of the sand plays a role; the coarser the sand, the longer the life span. Large beaches last longer than small replenished beaches. Other things being equal, a gravel-replenished beach should last much longer than a sand beach. Of course, if a big storm comes by with a certain intensity and from a certain direction, and with a long duration, all bets are off, and any replenished beach may disappear overnight in its first year.

Perhaps the longest life span of any artificial beach in the United States is that of Miami Beach, which has exceeded 20 years. Relatively minor erosion hot spot re-replenishment has been carried out there as needed, but it has been a functioning recreational beach and tourist attraction since it was first nourished in the early 1980s. The shortest life span may have been the Cape Hatteras, North Carolina, beach, which was pumped up in the 1970s. This beach disappeared in a storm even before the dredges finished their work.

After Hurricane Irene, the U.S. Army Corps of Engineers pumped in 280,000 cubic yards (about 214,000 cubic meters) of sand along a half-mile stretch (0.8 kilometers) of Atlantic City beach, New Jersey. Within five months most of it had disappeared, leaving an 18-foot beach scarp (sand cliff), the largest we have ever heard of. Scarps are far more common on replenished beaches than on natural beaches and reflect the fact that artificial beaches erode much more quickly than the natural beach they replaced. Plans are in place to pump in more sand along this stretch of New Jersey coastline at a cost of $4 million to get rid of the scarp.

It isn't just stretches of badly sited beachfront development that rely on replenishment for their survival. Beach replenishment is a vital component of Holland's sea defenses. Holland is in a position no country would wish to be in. For centuries, Holland has drained huge areas of tidal flats and converted them to agriculture. The country is now mostly below sea level and depends on a complex system of dikes and artificial dunes for its survival. Part of the defense strategy has been to maintain beaches in their position circa 1991 by beach replenishment. This was traditionally done in a pragmatic way by monitoring the volume of sand in the beach and replenishing sections that became too narrow. The average annual volume of replenishment sand in the 1990s was 8 million cubic yards (6 million cubic meters). Since 2000, it has increased to 16 million cubic yards (12 million cubic meters). Recent studies have suggested that the annual loss is actually 26 million cubic yards (20 million cubic meters) and so the volume of sand added will again have to be increased to maintain the coastline. Using an extremely conservative estimate of five inches (13 centimeters) of sea-level rise by 2100, Dutch engineers have calculated that the annual replenishment volume to hold the coast in its ideal position will be 111 million cubic yards (85 million cubic meters) by the year 2050.

Between March and October 2011 an enormous plug of sand was built north of Rotterdam Port by dredging the seabed. Known as the "sand engine," it is the brainchild of the ever-resourceful Dutch engineers charged with defending the country's coastline. In the perverse world of coastal engineering the project is billed as an example of "building with nature," when in reality it is an essential part of that country's never-ending war with nature. The single replenishment consisted of 28 million cubic yards (21.5 million cubic meters) of sand emplaced at a cost of US$96 million (€70 million). The idea is that the hook-shaped body of sand that extends one and a quarter miles (2 kilometers) into the sea will migrate along the shoreline, smearing the sand along the coast like an enormous paintbrush for five years, thus saving the costs of many smaller-scale replenishments and alleviating the growing demand for replenishment sand.

The significance of the never-ending process of beach replenishment is more than just an inconveniently high cost of replenishment after replenishment. The problem is that whatever damage is done (such as killing of all beach fauna) and whatever problems are caused (such as encouraging denser shorefront development), they are repeated with each replenishment and will go on and on and on.

A very frequent problem facing communities with replenished beaches is unrealistic estimates of their life spans. It is fair to say that the life span of almost every North American replenished beach has been overestimated. Up until the early 1990s, essentially all replenished beaches were given an expected life span of 10 years by the U.S. Army Corps of Engineers, a target that was hardly ever achieved.

A major exception to the rule has been the aforementioned Miami Beach. It is not clear why the beach at Miami Beach has lasted so long, but it may be because of the packing together of irregular shapes of the sand particles, all of which were calcareous remnants of marine organisms. Also, there is some patchy cementation of sand on the upper beach that might help hold it in place. On South Beach (the south end of Miami Beach), sand cementation is extensive enough to allow polo matches!

But the Miami experience has another side to it. The city has been forbidden from dredging sand from offshore again because previous dredging killed a lot of coral. The beach is purely calcium carbonate from various shells and fragments of reef coral. Such calcareous sand is soft and is easily ground up in the surf zone, producing a suspended cloud of fine fragments that can kill or drive away filter-feeding organisms.

More often than not, the erosion problem addressed by beach replenishment is one that is caused by humans through navigation-channel dredging, dam building on rivers cutting off beach-sand supply, and seawall and groin construction on the beach. Much of the problem of erosion and sand loss all over Florida is related to jetty construction and navigation-channel dredging.

A particularly blatant example of engineering incompetence occurred in 2004, when the U.S. Army Corps of Engineers removed an entire ebb tidal delta at Shallotte Inlet, North Carolina, to obtain sand for replenishing adjacent Ocean Isle Beach. The ebb tidal delta is an essential part of the barrier-island system. Sand transfers back and forth from one island to another across the inlet via the tidal delta. As should have been expected, the loss of the delta immediately led to greatly accelerated erosion downdrift on Ocean Isle Beach. The erosion there was so rapid that not enough sand could be dredged to fix it; instead, the dredging directly led to the construction of a sandbag seawall.

In Barrow, Alaska, the community mined its beach to provide sand to construct a new airstrip. A few years later, in 1999, the beach had to be replenished due to the "erosion" caused by the earlier mining project. The cost of sand was about $75 per cubic yard (0.8 cubic meters), roughly five to 10 times the cost in warmer latitudes.

Erosion caused by dredging, whether for replenishment or other purposes, is a problem that has been with us for a long time. On January 26, 1917, the tiny coastal village of Hallsands, England, was destroyed overnight in a big storm that struck at a spring tide. It was an early example of the complexity of coastal-engineering practice (a problem broadly discussed in the previous chapter). The cause was the mining of gravel from the nearshore seafloor that commenced in 1897 updrift from the town. This cut off the supply of gravel to the beach, which eventually led to a much-narrowed protective gravel beach in front of the village. Long before the village fell into the sea, local fishermen, intuitively recognizing what was happening to their vanishing beaches, managed to stop the dredging, but it was too late for Hallsands.

One common ploy of beach consultants is to compare the storm response of replenished and non-replenished beaches to demonstrate the value of replenished beaches. After a storm struck Myrtle Beach, South Carolina, with its new replenished beach, a consultant pointed out that little change had occurred on the beach, especially when compared to nearby much-damaged communities, such as Cherry Grove and North Myrtle Beach. What the consultant failed to notice (or say) was that the difference in the storm response in the other communities was because Cherry Grove and North Myrtle Beach are narrow, low-elevation islands without protective dunes, in contrast to Myrtle Beach, which has a higher elevation mainland shoreline. As always and in all things, consultant claims must be taken with caution.

There Is a Lack of Retrospection

After hundreds of individual episodes of beach replenishment on hundreds of beaches in the United States and elsewhere over 50 years, there still has been no systematic study of the history of replenishment projects, including beach life spans, and no careful examination of the design parameters from the mathematical models that have been used to predict beach behavior.

The mathematical models used to forecast such things as the life spans of replenished beaches and the environmental impacts of seawalls can only be accurate if one knows the future schedule of storms. Most important changes in replenished or natural beaches occur as a result of storms, and unless modelers, endowed with supernatural powers, know when and where the next storm will occur and how big it will be and from what direction it will come and how long it will last, accurate mathematical prediction of beach behavior is impossible.

The effect of this lack of retrospection, whether intended or not, is to allow the coastal-engineering community to continue its highly optimistic views of both beach life spans and environmental impacts. It also allows the continued and unfettered use of mathematical models, which give estimates of beach longevity an entirely unjustified aura of sophistication. When beaches fail to survive as predicted, a common excuse is that the beach disappeared because of an unexpected and unusual storm. Should any storm be unexpected on a beach?

In the U.S. the Army Corps of Engineers is primarily responsible for this lack of hindsight, but other agencies concerned with coastal management, such as the National Oceanic and Atmospheric Administration, are also reluctant to take on this hot potato. In the highly politically charged atmosphere of U.S. beach management, studies liable to cast doubts on engineering assumptions would not be appreciated.

Beach Replenishment Destroys the Ecosystem

The beach teems with life (see chapter 1) and is part of a connected ecosystem that extends from the dunes to well out on the continental shelf. The beach is an unusual ecosystem that adjusts to violent and rapid changes during storms. Some of the animals in beaches can actually detect early signs of impending storms and burrow to greater depths in the hopes of surviving the storm. But nothing prepares them for instant burial under tons of sand when the beach is replenished.

Immediately upon burial by beach-replenishment sand, the ecosystem disappears almost in its entirety. No foraging birds and crabs, no meiofauna to feed small crabs, no clams, nothing for fish to graze on, and, going up the chain, no mackerel and flounders offshore whose food has disappeared, and few sharks hanging around to devour the unwary fish. Replenishment makes the beaches resemble a small-scale movie version of a city abandoned as all the residents are killed by a mysterious toxic cloud. The toxic-cloud analogy is not far from reality, as mud in varying quantities is also released during replenishment, creating a sometimes-fatal problem for filter-feeding critters, including certain clams, sponges, and fish. Belgian ecologist Jeroen Speybroeck and 12 colleagues carried out a global review in 2006 of the ecological impact of beach replenishment and identified "sizeable impacts on several beach ecosystem components"—specifically, microphytobenthos (e.g., algae), vascular plants (e.g., creepers), terrestrial arthropods (e.g., ghost crabs), marine zoobenthos (e.g., clams), and avifauna (birds).

Even changes in beach shape brought about by replenishment can have important implications. A steep beach reduces the area of habitat for some species and usually receives more wave energy per square foot (0.1 square meters) than a gently sloping beach. Sand compaction, a common problem in artificial beaches, reduces permeability, water volumes, nutrients, and interstitial space for meiofauna.

How long does recovery take? A few studies have been made of this, the most notable (and notorious) of which was a $12 million U.S. Army Corps of Engineers analysis of the recovery of fauna on replenished beaches in New Jersey. Despite the huge cost of the study, the question of ecosystem impact was not answered, and the study has contributed little to the question at hand, a not-atypical result of studies by the Corps. Cynics suggest that the Corps used "dependable" consultants who would find the "correct" answers. In addition, the study used a long-ago-nourished beach as its "unspoiled natural beach" for comparison purposes. Most estimates indicate that within two to three years there is significant recovery of the ecosystem, but most artificial beaches are repeatedly replenished, thus starting the recovery process all over again.

Beach Replenishment Causes Damage to Offshore Ecosystems

The biological impacts of smothering a beach with sand are relatively obvious. The impact on the biota at the site of offshore sand mining is another matter: it's underwater, out of sight and out of mind, and it is a field almost devoid of meaningful research. The few studies made of the dredge-source sites indicate that "holes" remaining on the bottom do end up with a new fauna, sometimes dominated by marine worms. The holes, like any depression on the shallow seafloor, cause fine sediment to accumulate in the wave shelter of the depression. Normally, the mud and silt gradually move across the continental shelf into deeper water. In storms large enough for waves to stir up the seafloor, this accumulated mud is re-suspended, muddying the water, to the detriment of filter-feeding critters. The more mud in the hole and the longer the duration of the storm, the greater the extent of muddy waters.

An interesting example of this problem was a trench dug in the 1960s for a pipe to carry treated sewage offshore from the South Shore of Long Island, New York. The ditch was never used, and it rapidly accumulated mud in the years following its construction. For at least a decade, unusually muddy waters following storms were attributed to waves stirring up the accumulated mud in the ditch.

Dredging sand from offshore obviously kills all the organisms in the path of the wandering dredge head. Proof of this is offered by the seagulls swooping down to the beach to grab suddenly displaced creatures that were plucked up from the seafloor. How damaging this is depends on the nature of the offshore sand source. The 2011 Nags Head, North Carolina, replenishment obtained sand from a thick offshore sandbar, which meant that a relatively limited surface area was covered by the dredge. In contrast, a 2009 replenishment on Myrtle Beach, South Carolina, obtained sand by virtually vacuuming a large area, taking sand only from the upper few feet of the seafloor. The Myrtle Beach project must have been much more damaging to seafloor fauna than the Nags Head project, not only because it wiped out fauna and flora but also because natural seafloor habitats, such as surface irregularities, were altered. The commercial fishing community was largely responsible for halting a proposed replenishment project on Nantucket because of fears of damage to their livelihood. To our knowledge this is the only example in the United States of commercial fishers halting a major beach replenishment.

Dredges can cause damage to coral heads and coral reefs in more ways than one. Kicking up smothering fine sediment, as happened off Miami Beach, is a common problem. That is the reason why Miami is no longer allowed to use offshore sand in future replenishments. The same thing may have happened to coral heads off Maui's Sugar Cove. The sand was too fine, and wave action produced silt, which smothered reef coral. Off Boca Raton, Florida, a dredge dragged its giant anchor the length of the offshore reef, causing a great deal of harm.

On the Atlantic beaches of Florida, the "native" sand in south Florida was dominated by quartz, a hard mineral not likely to produce much suspended material in the surf zone. However, replenishment sand from offshore Florida is commonly a much softer calcareous material (shells and coral fragments) which gradually abrades in the surf zone. This produces a band of water containing fine suspended sediment parallel to the shoreline. The milky band of water is visible for tens of miles along the Florida shore and even from a commercial airliner coming in for a landing at Miami.

This silt- and clay-laden water kills coral heads, makes life difficult for all filter-feeding organisms, and, according to a study by geologist Harold Wanless, prevents the northward migration of coral larvae in response to warming waters caused by global climate change. The cloud of suspended sediment is like a giant wall in the way of migrating species.

None of this cuts any ice with the engineer concerned with finding a cheap source of sand. An interesting insight into the engineering mindset was provided from the dredging associated with building new wharves and undertaking the enormous "sand engine" beach replenishment near Rotterdam. A huge amount of sand was required, and the seabed was the obvious source, but what of the ecological impacts? Engineers were informed by ecologists that depressions on the seabed were ecologically richer than intervening high areas of the seabed. It was music to their ears, and the reasoning then went: if holes are good for the ecosystem, lots of dredged holes will improve it.

Beach Replenishment Causes Problems for Sea Turtles

The most visible of large warm-water animals that require the beach for some portion of their life cycles are the sea turtles. Seven species exist in waters off the eastern coast of North America; both coasts of Mexico and Central America; and the northern coast of South America. Of these, loggerheads, Kemp's Ridleys, green turtles, hawksbills, and leatherbacks are all considered endangered. Several international organizations assume that these turtles face a high risk of extinction.

The human-made hazards facing sea turtles are numerous. In ocean waters, dredges, fishing trawlers and long-liners, marine debris, and pollution are among the problems. On many occasions, dredges have killed turtles by sucking them into the pipes. The U.S. Fish and Wildlife Service has set a killing limit of six turtles, after which an operation must be shut down entirely. On Bogue Banks, one dredging operation for a replenishment project killed five turtles and tried everything possible to prevent the death of a sixth. The dredgers even had shrimp boats dragging nets on the seafloor around the dredge site to scare turtles away. The sixth turtle was finally caught in a pipe but miraculously came through to the beach, badly injured but still alive. The turtle was rushed to a turtle hospital on Topsail Island, where it was put on an IV drip and was patched up. The local mayor was cited as saying, "We're all rootin' for the little fella." Of course he was, because if the replenishment had been stopped, several houses were in immediate danger of collapse. Somehow the turtle survived and was released months later, somewhat worse for wear.

Erosion scarps, common features of replenished beaches, can be hazardous to turtles. Scarps (reflecting high rates of erosion) range in size from a few inches to 10 feet (from a few cm to 3 meters), as on Westhampton Beach, New York, and Wrightsville Beach, North Carolina, to the reportedly monumental 18-foot (5.5-meter) scarp on Atlantic City beach, New Jersey. A turtle heading up a beach must crawl over the scarp and, as a rule, a one-foot (0.3-meter) scarp is the highest that a turtle can scale.

The quality of replenishment sand is another critical factor for turtles. Some beaches are of such poor quality that nesting, at least in the first year or two after the beach is replenished, is impossible. On Isla Blanca Park, South Padre Island, Texas, and on Pine Knoll Shores, North Carolina, replenished beaches consisted of very muddy sand. Additionally, two replenished beaches on Emerald Isle, North Carolina, were shell hash—almost purely oyster shells with minor amounts of sand and mud.

A replenished beach on Oak Island, North Carolina, was laced with limestone cobbles and small boulders, which made turtle access and nest excavation difficult and dangerous. One unusual type of replenishment material trucked into Surf City, New Jersey, contained unexploded World War II ordnance. Dredging near Lewes, Delaware, turned up colonial-era artifacts from a long-ago sunken ship.

The thermal properties of sand are important in the incubation of turtle eggs. Color is partly responsible for beach-sand temperatures, and

FIGURE 31 A very bad replenished beach made up entirely of shell hash, on Bogue Banks, North Carolina. This resulted from lack of pre-project prospecting for high-quality sand and from poor enforcement of replenished sand-quality requirements. PHOTOGRAPH BY ORRIN PILKEY.

FIGURE 32 Erosion scarp on a replenished beach made up entirely of shell and coral fragments in the Maldives. Erosion scarps are a very common feature of replenished beaches, a reflection of the fact that replenished beaches erode very quickly relative to the natural beach they replaced. PHOTOGRAPH BY ANDREW COOPER.

the darker the sand, the warmer the sand, the higher the percentage of female hatchlings. Sand darker than the "native" beach sand is a characteristic of replenished beaches on the U.S. East Coast.

A 1995 review of the impacts of beach replenishment on turtles found that for nesting females "nourishment can cause (1) beach compaction, which can decrease nesting success, alter nest-chamber geometry, and alter nest concealment, and (2) escarpments, which can block turtles from reaching nesting areas." As far as the eggs and the hatchlings are concerned, the study found that replenishment could affect survival and development by "altering beach characteristics such as sand compaction, gaseous environment, hydric environment, contaminant levels, nutrient availability, and thermal environment." Replenishment in areas with incubating eggs could cause the eggs to be buried excessively deep.

THE UGLY

Beach replenishment, ironically, may be the single most important cause of beach loss in the future. Replenishment is a never-ending cycle of beach repair, which allows buildings to increase in size and height. Mom-and-pop beach cottages give way to high-rise condos, and the monetary value and political clout of the community both increase. The idea of moving buildings back or demolishing them in order to preserve the beach is rarely even a consideration. Sooner or later, as the sea level rises or as sand supplies give out, seawalls arrive on the scene and eventually lead to the demise of the beach.

Beach Replenishment Encourages Increased Density of Development

Development in Carolina Beach, North Carolina, changed from single-family beachfront development to multi-family and high-rise buildings almost immediately after a beach replenishment in 1982. This change was correctly attributed to the replenishment project both by the local media and by the local office (the Wilmington District) of the U.S. Army Corps of Engineers. On a larger scale, the giant replenishment of Miami Beach that same year was correctly given credit by the Corps for the increase in density of development of this community. There was no question that the replenished beaches in both of these communities led to intensification of development and to a greatly increased difficulty in responding to any future sea-level rise.

When it became apparent that increasing the density of development was considered a negative aspect of beach replenishment, the Corps completely reversed course and came out with a report (called the "Purple Report" after its cover color) in 1994 that purported to show that beach replenishment did not lead to increased development. It was an absurd conclusion. For example, before the 1982 Carolina Beach replenishment, which immediately preceded the aforementioned leap in development density, portions of the community were flooded on a frequent basis and locally no beach existed adjacent to a seawall. No sane developer would have invested there with such obvious indicators of a shoreline in trouble, but once the new beach arrived, there was no hesitation.

On Hilton Head Island, South Carolina, the setback line for development was moved in a seaward direction after a replenishment project. It was a move that ignored the fact that replenished beaches usually disappear much faster than natural beaches, and it pushed forward the date at which a beach-destroying seawall would need to be built.

Florida is the end point of the increasing density of development encouraged by beach replenishment. The hundreds of miles of high-rise-lined beaches there would not exist if not for beach replenishment. Wide beaches crowded with swimmers are almost certainly replenished beaches, especially in southeast Florida. To encourage the beach-driven tourist and second-home industry, high-rises were jammed against the shoreline, the closer the better. Why not? The beaches are wide, the government stays out of the way except to provide money to pump sand on the beaches, and everybody else is doing it!

But cheap sand for Florida's beaches is running out. According to the *New York Times*, Miami-Dade, Broward, and Palm Beach counties along the southeast corner of the state are almost out of sand on the adjacent continental shelf. Broward County is even considering resurrection of an old idea: grinding up recycled bottles to make sand.

As the sea level rises, replenished sand on the beaches will disappear at increasingly rapid rates—all this at a time when the federal government is clearly backing off future funding for artificial beaches. Seawalls are being built along more and more of Florida's shoreline reaches, and the federal government has expressed concern over the rate of seawall construction and its impact on turtle nesting. It doesn't take much imagination to see that decades down the road much of Florida will be beachless, seawalls will be high and mighty, and the main tourist pastime will be prome-nading atop the walls—all because beach replenishment provided a false sense of security.

Beach Replenishment Creates Swimming Hazards

Beaches have flavors. Some are used by families and others are used by fishers, nudists, surfers, gay people, drinkers, and so on. Throughout the world of beaches, the type of usage often depends on the nature of the beach, particularly the beach slope. The slope of the intertidal zone of beaches, including artificial ones, is controlled primarily by the grain size of the beach material: the coarser the sand, the steeper the beach.

Steep beaches are preferred by surfers because of the larger, steeper waves they produce. Families with small children of course prefer flat-ter beaches with small waves. A few years back, a family beach with fine sand, near Corpus Christi, Texas, was replenished with coarse sand. The result was a steepened beach, the drowning of a child, and a hue and cry to bring the beach back to what it was.

There were 32 spinal and neck injuries in 2009 and 11 in 2010 at Cape May, New Jersey, all caused by scarps on the replenished beach. Body surfers were most susceptible of all. At the end of a run, they sometimes crashed into the vertical sand wall of the scarp, with resulting neck in-juries. The reason these numbers came to light is that they were used as justification to get federal funding for more sand! The idea was to bury the scarp—a temporary solution if ever there was one. Better to have flat-tened the scarp with a bulldozer—a one-day, low-cost operation.

The most famous injury from a replenished-beach scarp was that of New Jersey governor Jim McGreevey, who broke his leg as he tumbled off a Cape May beach scarp while on a night stroll in 2002. Incidentally, McGreevey was one of three consecutive governors who broke their legs while in office. Jon Corzine broke his in a 2007 car crash and Christie Todd Whitman fell and broke her leg in a 1999 skiing accident. One website asked: does being a New Jersey governor predispose one to broken legs?

THE FUTURE

Like hard engineered structures (see chapter 3), much of the design of re-plenished beaches and their environmental impacts are based on mathe-matical models. In numerous court cases, the failings of models and the

FIGURE 33 A 10-plus-foot-high scarp (about 3 meters) on a replenished beach in Seaside Heights, New Jersey. Such scarps are clearly a hazard to beach users and make it difficult to walk to the beach. Former New Jersey governor Jim McGreevey fell off such a scarp and broke his leg. PHOTOGRAPH BY ANDREW COOPER.

uncertainties behind them have been pointed out to no avail. Judges are informed by agency or consulting engineers defending the projects that the models are sophisticated and state-of-the-art, and it's hard for a judge to get beyond that.

In 2008, a proposed beach-replenishment project, this one on Palm Beach, Florida, was opposed by the Surfrider Foundation. The question before administrative law judge Robert Meale was whether sand and mud movement from the beach would damage offshore coral. The project was rejected primarily because of the uncertainties in the mathematical model, as testified by geologist Robert Young, director of the Program for the Study of Developed Shorelines at Western Carolina University. The rejection, based on refutation of mathematical modeling, was a first in

North America to our knowledge, and we hope that it is an example that will be followed by other judges.

There are a number of approaches that could be taken to reduce the biological impact of replenishment and preserve the very important ecosystem of the beach environment. Retaining the same grain size of sand is essential to maintaining the original ecosystem after recovery. One approach used at Naples Beach, Florida, is screening the replenishment sand to get rid of large amounts of shell material. Beach ecologist Omar Defeo and colleagues suggest that replenishing shorter stretches of beaches or putting smaller volumes or thinner layers of sand on beaches at more-frequent intervals could help. Another approach that could work is to intersperse replenished beaches with areas that are not replenished. Presumably, with this approach, longshore currents will even out the shoreline at a slow-enough rate so that the beach fauna and flora will survive. This approach to replenishment is a favorite among surfers, who see in a crooked shoreline the potential for great waves.

The timing of replenishment can be important to beach ecosystems. In the United States, replenishment is forbidden during turtle-nesting seasons, a rule that is sometimes broken. Beaches elsewhere around the world may have a variety of other organisms that could benefit from timing restrictions on replenishment.

The American Shore and Beach Preservation Association (ASBPA) is the most visible and most powerful national organization promoting beach replenishment. They also are very much in the business of promoting beachfront development. Recognition by this group of the negative impacts of beach replenishment is non-existent, and its promotion of beach engineering is relentless and disingenuous.

In the ASBPA's January 2013 newsletter, *Coastal Voice*, the best-replenished-beach award for the southeastern United States was given to the Isle of Palms Beach in South Carolina, yet this beach was primarily shell hash, difficult to walk on, and a disaster for the ecosystem. It is a poor-quality beach. In the same issue, Presque Isle, Pennsylvania, was listed as the top park or habitat beach in the nation. As we discuss in chapter 3, the breakwaters in the beach restoration and replenishment project on Presque Isle are destroying the natural habitat that the designers promised to preserve.

Also in the same issue of *Coastal Voice*, the ASBPA notes that "for every dollar the federal government spends on beach nourishment, it gets an estimated $320 back in tax revenues." Taking this statement literally, if the estimate were true, beach replenishment could well be a means to reduce the nation's budget deficit. For example, the 21-mile-long (34-kilometer) beach replenishment in northern New Jersey that cost $13 million per mile (1.6 kilometers) should have provided more than $87 billion to the federal coffers!

THE END OF THE BEACH

Beach replenishment is now a virtually worldwide response to beach erosion. Rather than move, demolish, or let buildings collapse, communities prefer to add sand to the beach. Replenishment is said to be better than seawalls, but in the long view it is clear that replenishment will eventually lead to seawalls. The possibility of moving buildings back does not exist along much of the Florida peninsula because of the preponderance of unmovable high-rises. A generation from now, beachless beaches will be commonplace in Florida. And the same is true for heavily developed shorelines all over the world. Beach replenishment is not the answer—it is at best a Band-Aid.

There are many other reasons why beach replenishment is not the answer. It is certainly costly, unsustainable, and ecologically damaging, and it will become financially and technically impossible in times of rapidly rising sea levels. Far from being the answer, replenishment is simply forestalling the inevitable and creating the conditions for more and more development that will be put at risk. Despite the sales talk of those promoting (and benefiting from) beach replenishment (e.g., property owners and engineering companies), it is a desperate and shameful undertaking that says we are not prepared to allow the natural beach the space that it needs to respond to future storms and rising sea levels and thereby to save the next generation from having to bear the cost of response to sea-level rise.

THE PLASTISPHERE

Trash on the Beach

IN JUNE 2010, A NEW HOTEL opened in Rome to the cheers of a large and boisterous crowd. It was a most unusual hotel, consisting of only three rooms and two baths, with a long list of locals lining up to spend a night in it, free of charge. The Danish supermodel Helena Christensen was the inaugural guest. The cheering crowd was also most unusual, not exactly what one might expect at a hotel's opening ceremony. All were avid environmentalists concerned with the future of the world's beaches. The hotel stayed open for five days and was then dismantled to be rebuilt several times during a tour of European capitals. The hotel, the Corona Save the Beach Hotel, was constructed from 12 tons of trash collected on beaches all across Europe.

The Save the Beach organization, which sponsored the hotel, provided this somewhat boastful but admonishing message: "We have built the first hotel made of trash in the world. This will be the future of our holidays if we do noth-

FIGURE 34 The Corona Save the Beach Hotel, featuring three rooms and two baths, was designed by H. A. Schult and constructed from 12 tons of trash collected from European beaches. Here the hotel is in Madrid, a stop on its tour of European capitals. PHOTOGRAPH BY GIANLUCA BATTISTA. USED BY PERMISSION OF H. A. SCHULT AND SEÑOR GOLDWIN EVENTS AND NEWSMAKERS.

ing to preserve our beaches." The building was funded by the makers of Corona beer and designed by the German trash artist H. A. Schult, who stated: "The philosophy of this hotel is to expose the damage we are causing to the sea and the coastline."

Almost anything that is dumped in the ocean or in a river flowing to the ocean has the potential to end up on the beach. Some of this debris is natural, but the vast majority is a result of human activities. But unsightly and hazardous trash on beaches is just the tip of the iceberg. According to the Ocean Conservancy, more than a million seabirds are killed each year, along with 100,000 mammals and turtles, through interaction with trash in the ocean and on the beach.

Garbage in the ocean and on beaches affects organisms in many ways. The most critical effects are caused by entanglement and ingestion. Abandoned fishing nets are the most deadly source of entanglement, often capturing large animals, such as turtles, porpoises, and even whales. Once firmly attached to the animal, the nets cause cuts that may become fatally infected. Periodically, the news media recount biological expeditions to cut free the fishnets attached to whales. Many of these biological disasters are far out at sea and invisible to most of us, but similar problems are often evident on beaches as well. We have seen several instances of birds entangled with fishing lines; in one case, two dead pelicans were entangled together with such a line.

It isn't just the waste discarded by commercial fishers that causes problems. A couple walking on a Lake Ontario beach blogged that they found two dead cormorants next to a garbage can. The inference was that the birds had likely died from the effects of the garbage they were eating. The same blog bemoaned the fact that broken pieces of plastic toys littered the beaches, making it dangerous to walk barefoot.

Tragic memorials to the effects of marine trash are the bodies of juvenile (and adult) birds of various species found on beaches, their deteriorating bodies exposing plastic trash inside. A shocking set of pictures by Seattle-based photographer Chris Jordan of dead albatrosses, their stomachs filled with plastic debris, on beaches of the remote Midway Atoll in the Pacific Ocean, showed the global extent of the problem when published in August 2012. Adult birds, seeking food as they swoop over the ocean, mistake the plastic as something nutritious and eat it, or bring it back to their young in nests on the beach. The plastic particles provide a false sense of being full once swallowed, and as a result the birds (or other animals) starve to death.

Sometimes garbage particles kill birds by blocking their windpipes or digestive tracts. Plastic that makes it to the stomach displaces food space and lacerates the walls of the stomach. A study by Stephanie Avery-Gomm and five associates at the University of British Columbia determined that 58 percent of seabirds found dead on U.S. and Canadian North Pacific beaches in 1977 had ingested plastic. By 2009 that number had risen to 92 percent. And these numbers were prior to the arrival of the Japanese tsunami debris.

Scientist Mark Browne from University College Dublin in Ireland and his colleagues observed that microplastic particles (less than 0.04 inches or 1 millimeter in size) are found in all oceans, from the poles to the equator. They concluded that the principal source of these small particles, which are likely ingested by marine organisms, is sewage contaminated by polyester or acrylic fibers released when clothes are washed. In August 2012, Hong Kong beaches were covered in what looked to be a layer of snow, but it turned out to be nurdles (tiny plastic pellets used in the manufacture of plastic products). They came from containers that had been washed off ships in a nearby harbor. Nurdles are found worldwide in the ocean and on beaches. One study found that some remote Hawaiian beaches comprised over 70 percent nurdles. They are sometimes mistaken for fish eggs and eaten by predators. Aside from this, nurdles absorb chemicals including the contaminants DDE (dichlorodiphenyldichloroethylene) and PCB (polychlorinated biphenyl).

In a 2013 study, Erik Zettler of the Sea Education Association and colleagues from Woods Hole Oceanographic Institution characterized the accumulations of tiny bits of plastic debris in the ocean as "microbial reefs." They discovered a huge variety of microorganisms, including algae and 1,000 types of bacteria, as well as a whole assemblage of predators that feed on them and on fragments of floating plastic not much bigger than the head of a pin. The organisms were different from those in the surrounding waters, so the researchers dubbed the ecosystem the "plastisphere." Some of the microbes they noted were harmful ones, which brought up the question of the role of the plastisphere in transporting disease-causing microbes.

At the opposite end of the size scale, logs often wash up on beaches. This has always happened when trees collapse onto the beach or are carried down rivers, but commercial logging is driving the numbers up and up. For example, in the west African country of Gabon, logging is causing detrimental effects for both beaches and wildlife. Logs are floated on rivers to sawmills at the coast or are loaded onto ships, but some logs escape and float out to sea and end up on beaches. Besides cluttering the beaches, the logs interfere with the nesting of sea turtles by preventing them from crawling to the upper beach to lay their eggs. Hence, the turtles have to find another beach as close-by as possible to fulfill their nesting objective.

In Washington state, both in Puget Sound and along the open ocean, many beaches are covered by these escaped logs, which are distinguishable from "natural" logs by their chain-sawed flat ends. The logs actually retard the rates of shoreline retreat. Some beachfront dwellers have learned to their chagrin that when they "clean up" the beach by removing the logs, the rate of shoreline retreat suddenly increases.

FIGURE 35 A gabion seawall on Majuro, Marshall Islands, holding back the community garbage dump. Gabions are wire baskets filled with local stones or shells, and they usually last between five and seven years in saltwater. When this wall breaks up, local pollution will skyrocket. PHOTOGRAPH BY ORRIN PILKEY.

This cleanup of logs, brush, and trees in front of developed lots along bays and sounds is a global problem. The typical effect of log removal is an immediate jump in the shoreline-erosion rate. In such cases, a stiff price has been paid to make a beach "pretty," and very often the next step is to build a seawall. Another problem with log-covered open-ocean beaches is the unexpected movement of logs caused by the surf. This can trap and even kill unwary beach walkers clambering over the obstructions. Water skiers in British Columbia have to keep a constant lookout for escaped logs, as co-author Andrew Cooper unfortunately discovered firsthand.

The Ocean Conservancy supports an annual worldwide beach cleanup and keeps close records to determine sources and quantities of the various types of beach debris. In 2011, the 25th anniversary of the organization's founding, it published an informative summary of beach cleanups titled "Tracking Trash: 25 Years of Action for the Ocean."

Over those 25 years, volunteers from the organization collected 166,144,420 items that didn't belong on beaches. The largest category— more than 86 million items—could be described as picnic items, such as bottles and cans, food wrappers, and containers (the last being the most prevalent—14.7 million of them). The second-largest category (59.4 million) was litter associated with smoking, including 53 million cigarette butts. Fishing line, heavy rope, pallets, lobster traps, floats, and tarps from ocean-related activities, plus trash from commercial vessels, made up the third most common category, numbering 13 million items. "Dumping" items, such as tires, car parts, appliances, and 55-gallon drums, totaled 4.5 million. Last, at 2.5 million items, were medical and personal hygiene items, including 863,000 diapers.

The Ocean Conservancy's international breakdown of trash items on beaches indicates something of the nature of the people in a country—and perhaps also the thoroughness of the volunteer cleanup crews. While the largest numbers of smoking-related items (1.3 million) were found on the long U.S. shoreline, several other countries also had an abundance of these items. Those included Mexico, 57,000; Puerto Rico, 42,000; Dominican Republic, 26,500; Ecuador, 25,000; Malaysia, 19,500; Kenya, 19,000; Bangladesh, 16,000; and South Korea, 15,000. Most other countries recorded fewer than 2,000 smoking-related items on the beach. As for the picnic items collected, Mexico, the United States, Kenya, and Puerto Rico had particularly large numbers, each with more than 100,000 items.

There are many stories behind the accumulations of trash on beaches. In June of 2012, more than 50 syringes were washed up on Long Beach Island, New Jersey. The syringes arrived after a heavy rain and were believed to have come from storm drains, where street-drug users dumped their syringes. On Tybee Island, Georgia, an informal annual April gathering of young people, known as the Orange Crush, leaves behind a massive amount of beach garbage, according to local officials. The island's south beach is littered with cups, bottles, and even clothing after the event. On the uninhabited Masonboro Island, North Carolina, a single traditional Fourth of July event leaves tons of garbage on the beach. A tragedy of these common events, worldwide, is that everyone always assumes that someone else will clean up their mess.

Kuta Beach in Bali, recommended by Lonely Planet as one of the top places to visit, is often strewn with debris, reflecting a poor schedule of beach cleanup (and a poor beach-management program). It seems that the Kuta Beach trash is derived from nearby islands, not from local communities, and mainly comes ashore in the winter tourist season, when winds and tides are just right. Nai Harn Beach in Thailand and Sanya

Beach, China, suffer from the same problem, magnified by infrequent beach cleanups and a societal custom of leaving trash behind after a wonderful day at the beach.

In June of 2009, a five-year-old girl suffered severe burns to her feet when she stepped on a discarded disposable barbeque that had been buried in the sand by its inconsiderate former users at Ireland's Brittas Bay. It wasn't the first such incident, nor is it likely to be the last. As a child, Tina Aldatz Norris burned her feet on hot charcoals buried under the sand on a beach in Orange County, California. For months afterward the pain was unbearable, and even decades later, her feet are sensitive and prone to blistering. Inspired by her experiences, she established a company that creates designer insoles to help people walk more comfortably.

Mahahual, Mexico, a Caribbean beach community 40 miles (64 kilometers) north of the Belize border, has a particularly severe problem with beach trash. Evidently, converging ocean currents and winds bring garbage to this location, and the trash can be identified as being from both the Caribbean and Central America. Specific countries identified as trash sources include Cuba, Venezuela, Honduras, Brazil, Haiti, and Jamaica — all converging on Mahahual Beach.

The source of garbage on beaches near the city of Buenaventura, on the Pacific coast of Colombia, is the city itself. This city of 300,000 dumps some of its garbage at the head of Buenaventura Bay. Periodically, when the winds and the tides are just right, the trash floats 10 miles (16 kilometers) or more out to the mouth of the bay, depositing a thick blanket of cans, bottles, and plastic on the nearby barrier-island beaches. Similar to the situation at Kuta Beach and Mahahual, the resorts and communities near Buenaventura are not the source of their beach garbage. The trash comes from elsewhere.

Beach garbage will likely continue to increase on a global scale. In the United Kingdom, estimates of litter on beaches indicate an increase of 121 percent since 1994. Happily, at least one global source of beach garbage has been reduced to some extent. Ships used to consider the sea their own dump site, and everything no longer useful, whether liquid or solid, went over the stern. The International Maritime Organization's International Convention for the Prevention of Pollution from Ships (MARPOL), Annex V (1988), is an international agreement intended to prevent at-sea disposal of many forms of garbage. Ships' garbage is compacted into tight

FIGURE 36 On this busy Durrës, Albania, beach on the Adriatic Sea, sheep are rummaging through the garbage left behind by beachgoers. Piles of garbage on the back beach are not an uncommon scene in developing countries where there are no trash receptacles available. PHOTOGRAPH COURTESY OF GENT SHKULLAKU / GETTY IMAGES.

non-floating bundles and is supposed to be disposed of in port facilities; in some cases, it is incinerated onboard. It is clear from labels on some beach trash, however, that violations occur, but MARPOL seems to have significantly reduced the problem of disposal at sea.

THE BIG ONE

The great Japanese tsunami disaster of 2011 may have swept 5 million tons of debris out to sea. Much of it sank, but 1.5 million tons floated across the Pacific, following the whims of ocean currents and winds. It is the biggest single mass of floating trash ever contributed to the ocean at one time. The rubbish has already reached parts of the U.S. West Coast, and eventually the entire west coast of North America, from Alaska and the Aleutians to Baja California, is likely to receive some of it. Of course, ocean currents will ensure that some of the debris will land in Hawaii along the way. Much of the remaining flotsam and jetsam will join the Great Pacific

FIGURE 37 According to the National Oceanic and Atmospheric Administration (NOAA), "Kanapou Bay on the Island of Kaho'olawe in Hawaii is a hotspot for marine debris accumulation. Because of its remote location, removal is difficult, resulting in beaches that look more like a landfill." PHOTOGRAPH COURTESY OF NOAA MARINE DEBRIS PROGRAM.

Garbage Patch. A recent mathematical modeling effort in Japan has indicated that Alaska may receive much more debris than originally assumed.

Attached to some of the debris already found on U.S. shores is a species of barnacle that is considered a threat to local fisheries. A live fish native to Japanese waters was found in a barely afloat skiff that washed up on an Oregon beach two years after the tsunami. The fiberglass skiff also contained scallops, mussels, algae, crabs, and marine worms, while a 165-ton dock that landed in Long Beach, Washington, had 120 foreign species attached, raising serious questions about biosecurity.

Many other bulky objects are also afloat, including barges and house fragments, that may damage nearshore reefs as the debris drifts ashore. Large numbers of polystyrene foam buoys used by Japanese fishers and other objects such as Styrofoam packing pellets are slowly breaking up into tiny fragments that are attractive (and fatal) to some fish and birds.

As debris accumulates on beaches, most activities, including the resting, feeding, and nesting of birds and mammals, such as seals, will be severely affected.

National Public Radio in the United States recently focused attention on Montague Island, in Prince William Sound, of *Exxon Valdez* fame, where at least 40 tons of tsunami debris have washed ashore to date. A small organization, The Gulf of Alaska Keeper, is leading the cleanup effort, which closes down each year once cold weather sets in. The Gulf of Alaska Keeper has an office in Japan that raises funds from sympathetic Japanese individuals. The work on Montague Island is the first major cleanup along the extensive, torturous shoreline of Alaska and is largely a private affair. But without question, this eventually will have to become a government effort. Alaska politicians have asked politicians in Washington, D.C., for $45 million to clean up the shoreline, but it is not clear if the money will be forthcoming or even if $45 million will be sufficient. Most of the shoreline will have to be approached by boat, considerably adding to the cleanup cost.

With the toxic remnants of spilled oil, the Styrofoam particles, the invasive species, and the massive amounts of debris, there is a possibility of an ecosystem disaster along the shores of Alaska. Styrofoam, patented by Dow Chemical in 1944, is blue or white foamed polystyrene used just about everywhere for seemingly everything. Most commonly, one is likely to run into Styrofoam coffee cups, clamshells used to carry food home from a restaurant, and packing pellets that are frustratingly difficult to discard once a package is opened.

THE BOTTOM LINE

The volume of trash, like pollution, is increasing on beaches, and no end is in sight. Although beach cleanups are a widespread annual affair, only small progress can be claimed in reducing trash at its source. And of course every tsunami and big storm will significantly add to the problem. A variety of other initiatives have tried to address the issue of trash in different ways. Local authorities try various types of garbage bins, beach cleaning, and information campaigns to reduce littering on beaches by beachgoers. Some even argue that not providing bins is the best way to deal with it because bins allegedly attract litter.

Dealing with the distant garbage is the biggest challenge. Many coun-

FIGURE 38 A double whammy on a rocky beach along the Saudi Arabian coast. Here is a remnant of the Persian Gulf oil spill, the largest spill ever, combined with various types of trash. PHOTOGRAPH USED BY PERMISSION OF MILES HAYES AND JACQUELINE MICHEL.

tries have attempted to reduce plastic litter close to its source by introducing taxes or bans on plastic bags or encouraging recycling. There are countless education campaigns to try and stop littering, with many of them focusing on the ocean as the ultimate destination of garbage. Former New York mayor Michael Bloomberg proposed a ban on the commercial use of Styrofoam. Ecovative Design, a start-up company, claims to have invented a biodegradable substitute for Styrofoam made primarily from agricultural wastes, but its commercial production is still some way off. Meanwhile, Styrofoam continues to coat some New Jersey and New York beaches periodically just as a fresh snowfall would. Much of the Styrofoam floats off into the ocean, where its damage continues but remains invisible—until dead animals wash up on beaches.

While MARPOL has had an effect in reducing waste from ships, it cannot stop dumping at sea by unscrupulous operators. A global effort is needed to clean up the huge patches of plastic already in the central portions of all the oceans, enforce marine trash-disposal regulations, and have trained cleanup crews at the ready for the next trash crisis (in the fashion of oil-spill cleanup crews).

In the meantime, trash is another nail in the coffin for beaches as we know and love them. Increased amounts of trash lead to more (and more expensive) beach cleanups, with concomitant effects on the beach ecosystem. Also, as trash increases, beaches become less attractive places to visit—who would choose to swim or sunbathe in a trash dump?

EMMANUEL FARNACIO LEFT A NOTE behind when his vessel departed the Port of Vancouver, Washington, in September of 2000. Farnacio was a crew member of the Norwegian-flagged MV *Höegh Minerva*. He wrote in broken English that the vessel had a "magic pipe" that bypassed pollution-prevention equipment designed to separate oil and water. The magic pipe pumped oily bilge water directly into the ocean. Included in his note was a diagram showing the pipe's location and a plea: "Pls. Secreat [*sic*] for me. Somebody kill me about my information."

In May of 2001, an investigator from the Washington State Department of Ecology, alerted by Farnacio's note, boarded the vessel when it arrived in Tacoma. And sure enough, the pipe was there. The widely used term *magic pipe* refers to how waste oil will disappear like magic if you have the pipe.

It is illegal to dump oil at sea, according to the 1954 and 1973/78 international agreements known as OILPOL and

MARPOL, respectively (the International Convention for the Prevention of Pollution from Ships). Marine Defenders, an oil-spill educational group, believes that 85 to 90 percent of all vessels plying the seas comply with the international law but estimates that 5,000 to 7,500 large outlaw ships routinely dump somewhere between 70 million and 200 million gallons of oil annually. This is the largest source of "man-made" oil in the sea and is likely an important point of origin for the widespread small amounts of oil that wash up on beaches in the form of tar balls.

As it turned out, the *Höegh Minerva* had dumped hundreds of gallons of oily waste both in Puget Sound and at the mouth of the Columbia River. Höegh Fleet Services, which has 38 vessels, was fined US$3.5 million. Vincent Genovana, the second engineer of the offending vessel, was jailed for 30 days for concealing evidence of the magic pipe, and the whistle-blowing sailor, Farnacio, went on liberty from the ship US$300,000 richer.

Handwritten notes in struggling English are not uncommon. When the MT *Chem Faros* was boarded in March of 2010 in Morehead City, North Carolina, an oiler handed a Coast Guard inspector a note saying: "Good Morning Sir, I would like to let you know this ship discharging bilge illegally using by magic pipe. If you want to know illegal pipe there in the workshop five meters long with rubber." In this case, the vessel had dumped 13,200 gallons of oil-contaminated waste at sea 11 days before the inspector stepped aboard.

Paying whistle-blowing crew members up to half of the ship's fine has proven to be a successful method of entrapping guilty ships. Four crew members of the Greek-flagged vessel MV *Iorana* were paid US$125,000 each for revealing the crimes. In this case the note read: "We are asking help to any authorities concerned about this because we must protect our environment and our marine lives."

In the old days, oil was routinely dumped from most ships that plied the oceans. Today it is estimated that complying with MARPOL costs ship owners between US$30,000 and US$150,000 annually per ship, depending on the size of the vessel. If the cost is avoided by secretly dumping oil, the profit of seagoing vessels naturally increases. An added incentive for illegal dumping is that the waste-receiving facilities in some ports for offloading oil are difficult to use and difficult to find. Höegh Fleet Services was not the only company trying to avoid annual costs from handling oil properly. The shipping giant Evergreen International paid a whopping US$25 million fine in 2005 for dumping oil in the Pacific as well as the Atlantic. At least seven Evergreen vessels had magic pipes.

Orrin Pilkey, on a 1964 research cruise to the edge of the continental shelf off the coast of Georgia in the southeastern United States, came upon a giant oil tanker anchored in about 650 feet (200 meters) of water in the landward margin of the Gulf Stream. The rookie oceanographer was awed by the size of the tanker alongside the tiny 55-foot (17-meter) University of Georgia research vessel. As the research vessel steered around the tanker to get to deeper water, Pilkey spotted a long stream of brown liquid emanating from the stern of the vessel and flowing north with the clear blue water of the Gulf Stream. The vessel, 10 years after OILPOL was agreed upon, was cleaning its tanks and bilges before picking up a new load of oil. The prevailing attitude was that "dilution was the solution" and that the sea's huge volume of water would disperse pollutants to an acceptable level. Sometimes it is called over-the-horizon dumping. Today, cargo ships, as opposed to tankers, are more likely to be the spillers.

SOURCES OF OIL POLLUTION

In the following discussion, oil volumes are expressed mainly in gallons. The petroleum industry primarily reports volumes in barrels. Each barrel contains 42 gallons.

Sources of oil that ends up on beaches include shipwrecks (e.g., the *Amoco Cadiz*), blowouts from offshore wells (e.g., the Ixtoc I Mexican well), releases of various kinds caused by storms and storm surges (e.g., Hurricane Katrina), natural oil seeps (e.g., off the coast of Santa Barbara, California), and, strangest of all, purposeful spills for military purposes (e.g., the Gulf War).

Significant lengths of shoreline can be covered by individual oil spills. Thirteen hundred miles (2,100 kilometers) were affected by the 1989 *Exxon Valdez* shipwreck spill in Alaska; 1,100 miles (1,800 kilometers) by the 2010 Deepwater Horizon blowout in the Gulf of Mexico; and 502 miles (808 kilometers) of Persian Gulf beaches in the 1991 Gulf War. The latter spill, the largest in history, involved at least 400 million gallons deliberately released by Iraq to prevent an amphibious landing by U.S. Marines on the beaches of Kuwait.

Shipping Spills

The west coast of India, including Goa with its beach resorts, is often peppered with tar balls, particularly during the monsoon season. Following a particularly severe season in 2010, when beaches were covered with a few inches of tar balls, a chemical study to determine their origin showed that they originated from tankers carrying Southeast Asian crude oil. The ships were believed to have cleaned their tanks in the Arabian Sea, producing the tar balls that then washed up on the beaches.

Small spills are common on some beaches near the entrances of busy ports and oil terminals. Nearby beaches may have thin, patchy layers of oil covered by an equally thin layer of sand deposited there during fair weather conditions. The appearance of the beach surface gives no indication of the oil until the footprints of the first and most unlucky beach stroller of the day reveal the sticky problem underfoot. Most of the sticky beaches that have been reported in sites around the United States, including the Chesapeake Bay and on Galveston Island, probably disappear in subsequent storms, which return the oils to the offshore waters, where they turn into tar balls.

Often there are interesting human stories around oil tankers that despoil beaches. The 1978 *Amoco Cadiz* oil spill of 69 million gallons that blackened the Brittany beaches of France occurred while the captain of the tanker argued at length with the captain of a rescue tugboat about how much the rescue would cost, all the while ignoring the drift of the giant vessel. The *Amoco Cadiz* had broken its rudder during a storm, and the financial negotiations raged on for more than three hours, until the vessel ran up on rocks. The *Arco Anchorage*, while steaming to Port Angeles Harbor along the Strait of Juan de Fuca between Washington state and British Columbia, struck a rock and spilled 239,000 gallons of crude oil. This 1985 disaster occurred when the pilot, who had officially taken over command of the vessel, steered the ship into the rock over the objections of the captain, who knew the rock was there.

On the opposite side of the globe, the *Torrey Canyon* spill in 1967 occurred off the southwest coast of the United Kingdom. Apparently, the steering lever of the *Torrey Canyon* was on autopilot until the last minute and the helmsman, who was also working as a cook, was inexperienced. To compound things further, all charts on the vessel lacked sufficient

FIGURE 39 Heavily oiled pebble beach in La Coruña, Spain, in May 1976 after the TV *Urquiola* crude-oil spill of 21 million gallons. In all, 134 miles (about 215 kilometers) of shoreline were oiled. This spill provided the basis for the development of the Environmental Sensitivity Index, which ranks shorelines in terms of their sensitivity to oil spills and subsequent cleanup efforts. Note that the upper part of the beach is the most heavily oiled. On the mid-beach, the lines are swash marks formed as the tide retreats. PHOTOGRAPH USED BY PERMISSION OF MILES HAYES AND JACQUELINE MICHEL.

detail for navigation in shallow rocky waters. The ship struck a *reef* (a mariner's term for a submerged rock) about 15 miles (24 kilometers) west of Land's End, causing the largest spill ever at that time—32 million gallons of crude oil—which led to a slew of new regulations for tankers.

The *Exxon Valdez* spill in Prince William Sound, Alaska, resulted from an immense number of documented shortcomings. The captain was sleeping off a hard-drinking session, and the ship was under the command of the third mate. The crew was tired, reduced in number, and working 12- to 14-hour shifts. The crew of the *Valdez* was unaware that the U.S. Coast Guard was also operating with a reduced staff and was no longer tracking vessels in Prince William Sound. In addition, the Raytheon Collision Avoidance System on the vessel was not turned on. To make matters even worse, the port of Valdez had dismantled its oil-spill response team in 1981, cut back on its equipment, and designated untrained individuals from other agen-

FIGURE 40 Navy landing craft are being used in the *Exxon Valdez* cleanup on Smith Island in Prince William Sound, Alaska. Oil deeply penetrates into these boulder and cobble (grapefruit-size rocks) beaches, causing massive damage to the ecosystem. Because of the very large grain size of the beach components, this may have been the most damaging oil spill in history. PHOTOGRAPH COURTESY OF THE U.S. DEPARTMENT OF THE NAVY.

cies to be responders. As a result, drills had shown the Valdez response team to be woefully ill-trained and poorly equipped. In 1989 the most destructive oil spill ever known began as the *Exxon Valdez* struck a rock.

Estimates of the oil-spill volume from the *Exxon Valdez* are based on the difference in the oil volume in the vessel after the leak was stopped and the original volume that was pumped into the vessel at the terminal. No one knows for certain, however, the quantity of water that seeped into the ship as it was grounded on the rock.

Some oil spills just keep on going. In 1993, two tugboat-assisted barges collided at the entrance to Tampa Bay, Florida, resulting in an oil spill of 360,000 gallons of heavy fuel oil, gasoline, diesel, and jet fuel. Despite a large cleanup effort, long-term ecological damage resulted along local beaches, marshes, and mangrove swamps. All the locals breathed a sigh of

FIGURE 41 A grounded offshore oil rig that floated toward the beach off Dauphin Island, Alabama, after Hurricane Katrina in 2005. Hurricane Katrina caused at least 44 oil releases, large and small, some of which ended up on Gulf of Mexico beaches. PHOTOGRAPH USED BY PERMISSION OF ANDY COBURN, PROGRAM FOR THE STUDY OF DEVELOPED SHORELINES, WESTERN CAROLINA UNIVERSITY.

relief as cleanup and restoration efforts were completed, but their respite was short-lived. In 2000, the U.S. Army Corps of Engineers was dredging the nearby Blind Pass Inlet and the spill was suddenly active again. It turns out that pools of oil from the original spill had sunk to the inlet floor and were reactivated and refloated by the dredge.

Storms

Coastal floods, tsunamis, and storms can also cause oil spills. Inundation or flooding of pipelines, refineries, oil-storage facilities, and even gas stations can release oil. Two major storms caused spills that affected U.S. shorelines: Hurricane Katrina (2005), when there were 10 significant spills or releases of oil, and Superstorm Sandy (2012), which caused three significant spills.

Oil Seeps

There are many thousands of seafloor oil seeps around the world where oil flows through cracks, faults, and fissures in the seafloor and enters the ocean water column. In the early days of offshore oil exploration, drilling was often directed at seep sites. In the natural world, a wide range of microbes feed on the oil and break it down into other compounds. Generally, the microbes associated with a given natural oil seep exist in the right quantity to disperse much of it. In contrast, after a blowout from a well, the few microbes around are quickly overwhelmed and the oil spreads far and wide.

The largest natural oil seep in the world is in U.S. waters off Santa Barbara, where about 10,000 gallons of oil are released every day. Oil seeps exist in the Gulf of Mexico as well. During the time that the Deepwater Horizon rig was spewing oil, natural seeps in the Gulf of Mexico released somewhere between 2.5 million and 6.3 million gallons of oil, most of which was dispersed by microbes.

Oil in Sunken Vessels

There are many sunken vessels around the world, thousands of them sent to the bottom during wars. The oil remaining in their tanks poses a long-term hazard. Both the United States and the United Kingdom have wreck-oil removal programs, although funding of these efforts is neither certain nor secure. However, there are numerous examples of successful oil removal from wrecks many years after their original sinking, including:

- Removal of oil in 2003 from the SS *Jacob Luckenback*, sunk in 1953 off San Francisco.
- Removal of oil in 2006 from the SS *Catala*, grounded off Ocean Shores, Washington, in 1965.
- Removal of oil in 2009 from a Liberty ship, sunk in 1971 off Sabine Pass, Texas.
- No remaining oil was found in a 2011 investigation of the SS *Montebello*, sunk in 1941 off San Luis Obispo, California, loaded with 125 million gallons of oil and gas.

In the United States, the National Oceanic and Atmospheric Agency's Remediation of Underwater Legacy Environmental Threats (RULET) project identifies the location of potential sources of oil pollution from sunken vessels. Awareness of these vessels helps planning efforts and could aid states in the investigation of reported mystery spills—sightings of oil where a source is not immediately known or suspected.

Blowouts

Blowouts usually occur because of incompetent engineering or unanticipated high pressures of gas and oil (which is incompetent geological interpretation). The Ixtoc I well in Mexican waters in the Gulf of Mexico blew up and caught fire in September of 1979. The heavy mud that filled the well was supposed to hold back oil under pressure, but it failed, as did the blowout preventers. The blowout preventers were supposed to cut off and seal the drill pipe near the seafloor in such an emergency. During this incident, 139 million gallons of oil flowed out, of which 9.2 million gallons landed on Mexican beaches and 1.2 million gallons soiled the beaches of South Texas.

The British Petroleum (BP) Deepwater Horizon oil spill was the largest blowout spill in history (roughly 210 million gallons) and the second-largest oil spill ever. Three companies shared the blame: BP; Haliburton, a drilling-service corporation; and Transocean, the actual drilling company. A number of mistakes and shortcuts were documented, usually in the interest of cutting costs. All three of the companies involved released a great deal of misleading information during the 87 days of uncontained spilling. In addition, inattentive government regulators and inspectors share heavily in the blame.

Oil and War

Oil by the millions of gallons spilled into every ocean during World War II as combatants sank thousands of ships. At the start of America's participation in World War II, ship traffic along the U.S. East Coast was immediately attacked by German submarines. It was a naval disaster, perhaps the East Coast's mini-Pearl Harbor. Eighteen fully loaded oil tankers were sunk off North Carolina, where tankers moving south were forced into a fairly narrow and submarine-vulnerable path to avoid the north-flowing Gulf Stream and the tip of the Cape Hatteras shoals. The beaches of North Carolina were said to have been blackened for miles, but no obvious sign of the World War II oil remains today.

On January 23, 1991, during the first Gulf War, the occupying Iraqi forces released a massive amount of oil from a terminal in Kuwait. Smaller amounts of oil simultaneously emanated from some damaged tankers, oil terminals, and an Iraqi oil refinery. The whole purpose of the oil release, which probably amounted to 400 million gallons (although some estimates are as high as 520 million gallons), was to foil an expected invasion from the sea by U.S. forces. (For most spills, the estimates of oil volumes vary widely, often depending on the professional affiliation of the volume estimator.)

According to Miles Hayes and Jacqueline Michel of Research Planning, a consulting firm with a reputation for objectivity in politically hot issues such as oil spills (reported in an article by N. I. Tawfiq and D. A. Olsen), by 1993, 40 percent of the oil had evaporated, 10 percent was dissolved, 50 million gallons had been collected, and between 84 and 126 million gallons had washed ashore. As of 2002, there were 10.5 million cubic yards (8 million cubic meters) of oiled sediments: 45 percent on sheltered muddy tidal flats, 23 percent in salt marshes, 18.5 percent on sandy tidal flats, 11 percent on sandy beaches, and the rest in various minor habitats such as man-made structures. All told, this oil on the shoreline was about 10 percent of the original total of released oil.

Twenty-one years after the spill, most of the oil was gone from sandy intertidal zones (mostly removed by waves), except for patches of oil that solidified into asphalt on the upper beach. Marshes, mudflats, and sheltered beaches with less wave activity remain heavily impacted by the Gulf War oil.

TAR BALLS ON BEACHES

A portion of the oil that is dumped at sea ends up as tar balls on the beach. There are few swimmers or surfers who have not at one time or another stepped on a tar ball softened by the hot summer sun. If one searches hard enough, tar balls can be found on all beaches, but on some beaches one would have to be blind not to spot them. Tar balls form by both natural and human actions. Long before European colonization or the advent of petroleum exploration, the Karankawa people of Texas used natural tar balls collected from beaches to line their baskets and make them waterproof. That tar originated from natural seeps on the adjacent continen-

FIGURE 42 Here on Okaloosa Island at Fort Walton Beach on the Gulf Coast of Florida, a number of tar balls have washed ashore from the BP Deep Horizon oil spill. This big tar ball is an unusually large one. Most are the size of a pebble. This photograph was taken on June 16, 2010, a month before the well was capped. PHOTOGRAPH BY DREW BUCHANAN, IN PUBLIC DOMAIN ON WIKIPEDIA (HTTP://EN.WIKIPEDIA.ORG/WIKI/FILE:16POILSPILL.JPG).

tal shelf, but most tar balls today are the product of oil spills. Spilled crude oil initially floats on the sea as a slick, but it gradually gets broken into smaller pieces by wind and waves. The lighter parts of the oil evaporate away, leaving just the heavy tar. Once on the beach, the tar gradually breaks down through microbial action. Most tar balls are small, about the size of a coin, but they can be much bigger: a 3-foot-wide (1 meter), 100-pound tar ball washed up on Surfside Beach, Texas, in August 2013.

Tar balls sometimes wash up in such numbers that beaches have to be closed and cleaned. High concentrations can occur after large oil spills, seasonal changes in wind or wave patterns, and even after hurricanes. In Louisiana, Hurricane Isaac (2012) washed tons of tar balls ashore that likely had been accumulating on the seabed since the Deepwater Horizon oil spill in April of 2010.

Although they are mainly a nuisance to beachgoers, tar balls have also been reported to have harmful effects. Microbiologist Cova Arias found 10 times the expected level of *Vibrio vulnificus*, a leading cause of seafood-borne diseases, in tar balls washed up after the Deepwater Horizon incident. She warned beachgoers against handling the tar balls. Other studies have shown that some people develop skin rashes or have allergic reactions to the hydrocarbons found in crude oil.

Very likely, illegal dumping of oil is partly responsible for the widespread, routine occurrence of tar balls on beaches, the bane of beachfront tourist resorts, which end up with ruined carpets and towels and understandably unhappy patrons. Some resorts and public beaches even provide footbaths in the form of pans filled with a solvent that removes tar. MARPOL has definitely reduced the amount of tar on beaches from shipping sources, but there remains a correlation between tar abundance and the density of vessels plying nearshore waters. Mediterranean beaches, both European and North African, seem to have particularly widespread occurrences of tar balls, as do Gulf of Mexico beaches.

BEACH VULNERABILITY TO OIL

According to studies by geologist Miles Hayes and associates, the impact of spilled oil on shorelines is quite variable. The least vulnerable shoreline is a vertical rock cliff. The most vulnerable are salt marshes and mangrove swamps. Mangroves along the margin of the Niger River Delta in west Africa have been subjected to so many minor oil spills that mangrove wood sometimes explodes when used in cooking fires.

Beaches occupy a vulnerability position somewhere between cliffs and marshes. Penetration of oil into the beach sand depends on several factors, including grain size: the larger the grains, the more the penetration. Maximum penetration of several feet or more occurs on boulder beaches, such as those in Prince William Sound. High-latitude beaches where glaciers once occurred are commonly gravel and hence are particularly susceptible to oil spills. The same can be said for beaches in warm water where coral reefs occur. Such beaches are often made up of coarse fragments of coral that allow the deep penetration of oil.

Maximum oil penetration in fine-sand beaches (grains barely visible to the naked eye) is around one-half inch (1.3 centimeters). In coarse sand (the size of sugar grains), penetration can be as much as 10 inches (25 centimeters). Bubbly sand, the soft sand you sometimes encounter on a beach walk, forms when incoming tides push air up through the sand, forming cavities below the beach surface. These cavities allow the rapid penetration of relatively large volumes of oil to the depth of the bubbles (4 to 20 inches or 10 to 50 centimeters) and beyond. Deep ghost-crab and ghost-shrimp burrows add to the depth-of-penetration predicament.

Bubbly sand is best developed on fine-sand beaches with a high tidal amplitude. In the United States, on Georgia beaches, bubbly sand is thick (up to about 10 inches or 50 centimeters) and widespread, the sand is fine, and the tide range is 7 to 11 feet (about 2 to 3 meters). On Cape Hatteras beaches in North Carolina, the tide range is less than 3 feet (0.9 meters), and the sand is coarse. As a result, the bubbly sand layer is thinner there. The bottom line is that an oil spill on Georgia beaches would do much more damage to the ecosystem (because of more air bubbles) than one on Cape Hatteras (even though it has coarser sand).

The amount of environmental damage done and the success of the cleanup vary with the type of oil spilled. Gasoline, the lightest petroleum product, usually evaporates within a few days and cannot be cleaned up. There are high percentages of toxic compounds in light oils that are damaging to the biota of the water column and the beach. Heavy oil, such as Bunker C fuel oil (a common type that is spilled), doesn't evaporate significantly and weathers very slowly. It is very damaging to the quality of beaches and to the biota and is extremely difficult to clean up. The crude oil from the Alberta Tar Sands presents a different problem. It is heavy and will sink to the seafloor (or lake or river bed). The technology for the seafloor cleanup of a big sunken spill has not yet been developed.

SO WHAT?

"So what?" is always a good question to ask about any study in order to put it in the context of the real world. So why worry about oil spilled in the nearshore?

The 1989 *Exxon Valdez* spill, amounting to between 11 and 33 million gallons, depending on whose estimate you accept, may have been the most devastating oil spill in history and was one of the most destructive human-caused environmental disasters ever. The reason for the magnitude of this disaster was primarily the rocky nature of most of the affected shorelines. Oil seeps deep into the spaces between cobbles and boulders. After the spill, much of the shoreline, including some sand beaches, was steam cleaned, which is also a biologically devastating process. Dispersants and solvents were widely used, and one partially successful attempt was made to burn off the oil. More than 11,000 workers were employed in the cleanup. A pristine, seldom-visited wilderness was turned into a giant campground. Hitherto-unknown archaeological sites were discovered and ransacked by some of the workers. The cleanup cost Exxon about US$2 billion.

Overall, the *Exxon Valdez* spill and its cleanup basically destroyed the entire shoreline ecosystem. Among the larger components of the system, more than 200,000 seabirds, 250 bald eagles, 2,800 sea otters, 300 harbor seals, and at least 22 orca whales were killed. On any oiled beach, bird nesting and turtle nesting are halted, as is clamming, both commercial and recreational.

As for human impacts in the case of a large spill, unless massive and costly cleanup efforts are carried out, the oil always closes beaches for years. After the 2010 BP spill, the cleanup removed much of the beached oil and quickly made many beaches usable once again by tourists.

University of New Hampshire biologists sampled beaches on Dauphin Island, Alabama, and Grand Isle, Louisiana, a few days after the Deepwater Horizon blowout, long before the oil reached the beach. Four months later they re-sampled the beaches and found major changes. The

FIGURE 43 An oil-spill cleanup on the tourist beach Ao Phrao in Ko Samet, Thailand. The oil originated from a pipeline leak on an offshore-drilling platform in July 2013. PHOTOGRAPH COURTESY OF ROENGRIT KONGMUANG/GREENPEACE.

before-spill samples contained the usual highly diverse assemblage of microbial organisms (the foundation of the beach food chain), including bacteria, nematodes, copepods, and protists. In 2012, two years after the spill, fungi dominated the living assemblage—despite the fact that the appearance of the beaches gave no hint that they had been impacted by an oil spill and a spill cleanup program. Obviously, important biological impacts linger on from a spill even after the disappearance of visible surface oil on the beach and in the region.

HOW TO KEEP OIL OFF BEACHES

Much of the oil that ends up on beaches comes from relatively small spills at inlets, navigation channels, and approaches to port facilities. Ideally, ports and harbors, large and small, should be prepared to react very quickly to spills ranging in size from bucketfuls to tons. They should also have the facilities to receive waste at low cost from commercial vessels. Pre-booming—putting oil-spill booms in place when vessels receive oil—

is critical and is required by some states, such as Washington. This is not a federal requirement in the United States, and there is little justification as to why not. Furthermore, companies that transport oil should always have their vessels accompanied by tugboats as they approach and enter the harbor or navigate through narrow channels. Spill-response equipment should be pre-staged in the harbor for quick response.

However, there will always be accidents, tanker groundings, and well failures, and for the foreseeable future, they will likely increase as both our consumption of oil and production of oil offshore increase. We can also expect that government regulations and enforcement of regulations will remain problematic at best. For example, solving the tar-ball problems on Goa's beaches could only be done with the rigid enforcement of international law. Oil on beaches will continue to be a problem, not to the levels seen during and after World War II, but it will remain a beach-quality factor for the foreseeable future.

STUCK IN A RUT

Driving on the Beach

WHAT DO MUZHAPPILANGAD BEACH, India; Kuakata, Bangladesh; Biakpai Beach, Nigeria; Chidenguele, Mozambique; Necochea, Argentina; Chirihama Beach, Japan; Dor Beach, Israel; Portstewart Strand, Northern Ireland; Rømø Beach, Denmark; Long Beach, New Jersey; and Long Beach, Washington, have in common? The answer: they are all beaches where driving a car is allowed. Well, not quite. Most beach driving in Israel and Mozambique is illegal, but it goes on nonetheless. Up to 3,000 cars can be seen on Portstewart Strand on a fine summer afternoon. At Southport, England, and on Assateague Island, Virginia, parts of the beach have been designated as car parks with capacity in the hundreds to accommodate beach visitors. On many other beaches, even in national parks where driving is strictly prohibited, "official" vehicles are allowed to patrol up and down the beach.

Along densely forested or rugged rocky coasts, beaches have served as convenient and easy routes for people to get from one place to another since prehistoric times. In some places beaches continue to serve as roads, even though our modes of transport have changed. Before cars, it was horses, and before horses, people walked. A 15-year-old boy, John Ross, walked about 375 miles (600 kilometers) through South Africa to Mozambique—from Port Natal (now Durban) to Delagoa Bay (now Maputo)—in 1827, to seek medical supplies for the fledgling British colony. Most of his walk was along the beach, not only because it was safer but, aside from a few rivers, it was an unobstructed route.

Inevitably, the long, flat surface of some beaches that were used as natural highways and byways gave rise in time to recreational activities. Horse racing at the shoreline became a popular pastime by the mid-nineteenth century. In 1845 the Sanlúcar de Barrameda Beach horse race was established in Spain, and in Laytown, Ireland, horse racing on the beach began in 1876. In time, motorized vehicles followed, and not long after the speed of horseless carriages was first measured in 1903 on Daytona Beach, Florida, auto racing became the rage.

The 7-mile-long (11 kilometers) Pendine Sands beach in Wales was the location for a momentous event in 1924. Driving on the beach at the then-amazing speed of 146.16 mph (235.22 kph), Sir Malcolm Campbell broke the land-speed record in a 350 hp Sunbeam V12. Beaches were apparently the location of choice for early attempts at the land-speed record, and Campbell went on to break eight more records until 1935—three of them at Pendine Sands and five at Daytona Beach. A rival speedster, J. G. Parry-Thomas, driving a vehicle that he had built known as "Babs," set a new record of 169.3 mph (272.46 kph) on April 27, 1926, which he broke the following day when he reached 171 mph (275.2 kph).

Competition with Campbell was fierce, and in an attempt to set a new record in March 1927, Parry-Thomas was killed. His crew buried the car on Pendine Sands, where it lay until it was discovered and exhumed in 1969. Although attempts at the land-speed record now usually take place in desert salt pans, Pendine Sands continues to attract fast drivers. In June of 2000, Don Wales, Campbell's grandson, set the United Kingdom's land-speed record for electric cars (137 mph; 220.48 kph). In recognition of these momentous events, the Museum of Speed is located in the village of Pendine.

Ultimately, however, driving on beaches was bound to become a problem for other beachgoers. The incompatibility of beach pedestrians and fast cars resulted in a ban on driving on Pendine Sands between 2004 and 2010, after a report by the Royal Society for the Prevention of Accidents warned of the danger of fatal accidents. The BBC reported on as many as 30 cars racing up and down the beach causing a serious hazard to beachgoers, even after the ban. However, a managed parking lot was reinstated at the beach on weekends and public holidays in 2010, because fewer visitors were coming to the resort.

It isn't only cars that achieved great speeds on Pendine Sands. In 1933, Amy Johnson and Jim Millison took off from the same beach in an attempt to fly nonstop to New York. They crossed the Atlantic but were blown off course and landed instead in Bridgeport, Connecticut.

Use of beaches for landings and takeoffs of light aircraft continues to this day, but the most famous landing was made by Charles Lindbergh in 1927. He landed the *Spirit of St. Louis* on Old Orchard Beach, Maine, because the local airport at Portland was fogged in. Currently, on the Scottish island of Barra, the official runway for scheduled airline flights is located on the wide sandy beach, and the aircraft have been specially modified for landing. When the airstrip was first established, a deal had to be struck with the local cockle fishers (clammers) who agreed not to dig on the runway.

Aircraft, however, are not terribly common on beaches and do not pose a serious threat. Cars and trucks, on the other hand, are such an issue on beaches around the world that they are increasingly being banned or restricted. In 2001, when almost all beach driving was prohibited on South African beaches, the positive effect on the ecosystem was immediate. The beach fauna and flora returned and the numbers and species of birds immediately increased. Most interesting of all was the return of the leopards. Prior to the coming of automobile traffic, leopards occasionally meandered along remote beaches before returning to their more traditional habitats. Now they have again been spotted on some beaches, to the delight of wildlife lovers.

The transition from driving to non-driving in South Africa was reasonably peaceful, although political pressure remains in that country to rescind or alter the ban. Indeed, permits can now be sought for a variety of beach events, including fishing competitions and boat launchings, and disabled people can also apply for permits to drive onto beaches that are otherwise inaccessible.

FIGURE 44 Fun on the beach at the annual kite festival in Rømø, Denmark.
PHOTOGRAPH BY ANDREW COOPER.

On the Outer Banks in North Carolina, the transition to car-free beaches has been rather less smooth. Residents are furious that the National Park Service is restricting driving along some shoreline reaches to facilitate survival of piping plovers and other beach-nesting birds, as well as turtle nests. The depth of the ugliness in the beach-driving controversy is reflected by a sign along coastal Highway 12 near Cape Hatteras. It reads "The Audubon Society—American Terrorists"! A representative of the society living in the town of Buxton on the Cape has had nails thrown in his driveway and "wanted" posters, featuring the image of his house, tacked to trees. Ugly indeed, and a measure of the difficulty of changing long-ingrained habits on beaches.

Many people feel that it is their God-given right to drive on the beach and bitterly resent it when this privilege is removed. In some parts of the world, the beach used to be the main road, and this is still true in some places, particularly in the developing world, and in remote locations in Australia and New Zealand. In the United States, residents of Fire Island, New York, drive along the beach to reach the causeway to Long Island. However, most beach driving is undertaken for recreation.

There's no doubt about it. Driving on a beach can be a very exciting experience. It can also be bumpy when cross-beach channels, sandbars, or abandoned sand castles are encountered, and it can be frustrating or even dangerous when the sand is soft. Hundreds of vehicles have been trapped in soft sand, only to be submerged by the next high tide. The 124-mile (200-kilometer) stretch of beach in northern KwaZulu-Natal, South Africa, was a veritable 4x4-vehicle graveyard before the country's beach-driving ban came into force. At the north side of Oregon Inlet on the Outer Banks, one vehicle's location, after several tidal cycles, was marked only by the tip of its antenna protruding from the sand.

THE IMPACT OF BEACH DRIVING

Driving on the beach is not a benign pastime. It affects the beach in many ways, some of which are not immediately obvious. Some of the issues arise between drivers and other beachgoers, while some relate to the beach itself and the creatures that live there. Some of the main impacts are discussed below.

Aesthetics of the Beach

Perhaps the most obvious annoyance to beach visitors is the sight of the closely spaced ruts extending down the beach as far as the eye can see. It is a most disconcerting experience for those who view beaches as a priceless piece of nature at the junction of the sea and the land. Thomas Schlacher and colleagues, in a 2008 article about Flinders and Main Beaches on North Stradbroke Island, Australia, showed that between 54 and 61 percent of both beaches was covered by tire tracks, with a maximum of 90 percent coverage. At Umhlanga Rocks, a popular beach north of Durban, South Africa, every morning at 8:15 a.m., the beach is traversed by lifeguards on a 4x4 vehicle. Later in the day, municipal vehicles drive along the beach to empty trash cans, and periodically police vehicles cruise along the beach. Despite a general ban on beach driving, it is almost impossible to see this beach without tire tracks on it (the municipal services are exempt from the ban).

FIGURE 45 Tire tracks on soft beaches, such as this scene on Bogue Banks, North Carolina, are particularly disastrous both for the beach fauna and elderly people's access to the water. PHOTOGRAPH USED BY PERMISSION OF DIANE NIES AND GREG NIES.

Restriction of Beach Use

On temperate beaches the car often serves a multitude of purposes for beach visitors. It delivers them to the beach and serves as a central location for picnicking, sunbathing, and changing into and out of bathing gear. It also provides shelter from wind or rain. All of those benefits, however, come at a price: the cars pose a risk to other beach users. Children running between cars are at risk of being knocked down, and without proper road markings, the cars also are a danger to each other.

On Emerald Isle, North Carolina, where unrestricted driving is allowed except in the summer, aging retirees complain that tire ruts, sometimes as much as 1 foot (0.3 meters) deep on soft sand beaches, prevent their access to the beach. In addition, beach driving on heavily used hard beaches, such as Daytona Beach, can significantly reduce the area of beach that can be used for sunbathing and other activities by non-drivers. The safety of beach users is a major concern to operators of beaches that permit cars. One online commentator reckoned that "you are more likely to get knocked down on the [Daytona] beach than crossing the road to get to it."

Damage to the Ecosystem

The beach is a complex ecosystem, from the microscopic meiofauna that live between sand grains to the various crabs and clams and the fish and birds that feed on them, to the Great White shark cruising just offshore. As the classic 1942 study by A. S. Pearse notes, beaches contain enormous food resources that sustain the rich fauna. The aforementioned Australian study of North Stradbroke Island by Thomas Schlacher and colleagues indicated that between 6 percent and 10 percent of available faunal-habitat sand matrix *was disrupted each day* by beach driving. The study also estimated that on a single day beach traffic disrupted 50,000 cubic yards (38,000 cubic meters) of sand on Main Beach and 16,500 cubic yards (12,570 cubic meters) on the shorter Flinders Beach.

At high tides, and especially at spring tides, off-road vehicles are often forced to the upper beach, where they have little space to avoid bird and turtle nests and a variety of other organisms. Driving over wrack lines made up of seaweed (an important source of beach nutrients) can be very damaging to the often-extensive fauna and flora clinging to and finding shelter in the wrack.

On Washington state's Long Beach peninsula, preservation of razor clams, the harvest of which is a major beach activity in season, is front and center in beach-driving management. In Brazil, some anchovy species and herring spawn on driveable beaches. On the Delmarva Peninsula on the U.S. East Coast, fish such as pompano and rough silvers depend on the beach. Southern California has the grunions, small, edible fish that are 4 to 6 inches (about 10 to 15 centimeters) long and celebrated in songs such as Frank Zappa's "Grunion Run" and Sandra Loh's "Night of the Grunion." Grunion spawning takes place on beaches when the fish rush ashore at the highest spring tides and lay and fertilize their eggs. Both beach driving and beach cleaning destroy the grunion eggs, so restrictions are needed until the eggs hatch, two weeks after the spawning.

Maybe the most graphic illustration of the effects of vehicles on the beach ecosystem is the effect on creatures that want to cross the beach but are prevented from doing so by deep tire tracks. Ghost crabs and especially baby turtles fall into these ruts, cannot climb out, and are doomed to crawl along the rut until they find a shallower rut or they die. Studies have shown that even ruts as shallow as 6 inches (15 centimeters) trap almost all turtle hatchlings. Several studies document the impact on ghost crabs, showing that they are particularly vulnerable when they emerge from their burrows at dusk. Ghost crabs are killed in large numbers on the beach surface by passing vehicles. In fact, Schlacher and associates calculated that 10 vehicles in one day could kill 7.5 percent of the ghost crab population.

Increased Erosion Rates?

Vehicle use on sandy beaches may affect sediments by altering their characteristics (e.g., grain size, organic matter, and moisture content), increasing compaction, and mobilizing surface layers, potentially leading to changes in the natural profiles of beaches. This is an aspect of beach driving that needs more research, but it appears likely that loosening of the surface sand and the formation of a rutted irregular surface make it easier for both wind and swash to pick up and move individual sand grains. Tire ruts may form sediment traps that reduce sand contribution from the beach to the dunes. In addition, destruction of plants and plant roots reduces their role in stabilizing sand, particularly on the foredunes. A study on Fire Island, New York, showed a significant loss of foredune vegetation caused by off-road vehicle traffic. This in turn caused increased rates of erosion during storms.

Car Crashes

Stories of deaths and injuries from car crashes on beaches abound. The driving record on the ocean beach of Fraser Island, Australia, must be among the worst in the world (right up there with Daytona Beach). Between 2003 and 2009, more than 40 serious accidents occurred, resulting in more than 100 injuries and several deaths. One Japanese driver of a vehicle that overturned and killed another tourist explained that he had watched the required safety video before starting out but didn't understand it because it was in English.

Pedestrian Deaths

Stories of run-over sunbathers on beaches all over the world also abound. Two drunk government bodyguards in Antioquia, Colombia, drove recklessly onto a beach and ran over three sunbathers, all of whom miraculously survived. On Jacksonville Beach, Florida, a police SUV ran over and seriously injured a sunbathing woman. In 2010, two 4-year-olds were run over and killed on Daytona Beach, in separate accidents. Both kids dashed out into the path of the cars. In 2012, the Daytona Beach city commissioners stopped all night driving on the beach, and calls from concerned citizens continue to urge a stop to *all* beach driving. Clearly, beach driving and sunbathing don't mix.

TYPES OF BEACH-DRIVING REGULATIONS

There are many ways of restricting driving on beaches short of complete prohibition, as in South Africa. Listed here are a number of approaches taken by governments at various levels. The success of these, of course, depends on the degree of enforcement, which, not surprisingly, varies widely.

- Requiring licenses, permits, or passes for a price: New Jersey
- Enforcing speed limits: Daytona Beach, Florida
- Specifying beach-access ramps: Oregon
- Specifying beach-driving zones: most beaches around the world
- Prohibiting dune driving: most beaches around the world
- Seasonal driving depending on turtles and birds: Emerald Isle, on Bogue Banks, North Carolina
- Seasonal driving depending on the level of pedestrian use: Bogue Banks, North Carolina
- Requiring certain vehicle types: Australia
- Complete prohibition of beach driving: South Africa and Israel
- Charging a fee as vehicles enter the beach: Portstewart, Northern Ireland

California

California allows almost no beach driving by the public. But still a lot of driving occurs by officials, such as police patrols, lifeguard trucks, and various kinds of maintenance and beach-cleanup vehicles. There is probably not a coastal tourist town in the world that depends on the beach for its livelihood that does not have a lot of *official* driving activity. Prohibition of beach driving in California requires extensive upland parking facilities.

Georgia

All of Georgia's barrier islands are off-limits to driving except for Cumberland Island, a national seashore at the southern end of the state's shoreline. The beach there has the largest number of turtle nests in the state. Ironically, the public cannot drive there, but the national seashore issues 350 driving permits, mostly to friends and families of landowners who live within the park boundaries, as part of the agreement that was made when the park was established.

Delaware

Delaware has unusual rules for beach driving. Only fishers are allowed vehicular access. Each vehicle must have fishing gear, and occupants can only engage in fishing activities.

North Carolina

North Carolina has a long tradition of beach driving on much of its 350-mile shoreline (about 560 kilometers). When the federal government wanted to establish the Cape Lookout National Seashore on the southern half of the Outer Banks (Core Banks and Shackleford Banks), it was prevented from doing so by the beach drivers who had a powerful senator in their corner.

At the time, the 21-mile-long (34 kilometers) Core Banks barrier island, which had no permanent residents, was a virtual graveyard of abandoned cars and piles of beer cans and wine and liquor bottles left behind by generations of fishers. The custom was to bring over a car and drive it until it stopped running and leave it right there, marking the moment of death. One abandoned school bus located at a commonly used campsite was completely filled from back to front and to the roof with beer cans.

Finally a compromise was reached with North Carolina senator Jesse Helms, who allowed the park plans to proceed, provided that 800 cars were allowed to remain on the island. The National Park Service gained a partial victory when it required that all vehicles be inspected annually, which means that each of the 800 vehicles has to be transported each year to the mainland via barges. After the National Park Service took over, all the abandoned cars were gathered together and temporarily stored in a spectacular pile of 2,000 cars waiting to be carried by barges to a mainland junk dealer.

The second island in the Cape Lookout National Seashore, Shackleford Banks, had no tradition of beach driving, but unfortunately the National Park Service began a daily transit on the beach along the entire island. It is a means of checking up on things, but do wilderness areas really need daily scrutiny? When the park was first formed, patrolling was done on foot, and the shells on this extraordinary shelly beach remained uncrushed by official tires. That is no longer the case.

Texas

The state of Texas has long had the Open Beaches Act, which allows unfettered vehicular access to most of the state's beaches, except for certain areas on the Padre Island National Seashore and parts of western Galveston Island. Shoreline retreat is occurring on almost all of the western part of Galveston Island, and rates of retreat will increase in the future as the sea level rises. Driving on the beach will likely become a thing of the past.

A recent court case changed the way beach driving is viewed in Texas, angering some drivers. It came about when an individual who owned several houses that ended up out on the beach as the shoreline retreated past them during a storm initiated a lawsuit. The state ordered the removal of the houses so that driving could continue unabated and even offered to pay the removal costs. But a higher court reversed the decision and allowed the houses to remain. The court stated that even though vehicular access was unrestricted before a storm, once a shoreline moved back, access no longer existed on the beach that was newly formed by shoreline retreat. Now, vehicles are allowed on western Galveston Island as long as they avoid the "new" beach. Putting it in another light, the state's legal system (like those in most states) is unable to cope with the extreme dynamics of the shoreline in a time of rising sea levels.

Adding to the ire of local beach drivers is the fact that Carol Severance, who instituted the fateful lawsuit, is a California resident!

Oregon

In 1913, the State of Oregon declared that its entire shoreline was a public highway. Eventually, 35 parks with beach accesses 10 miles (16 kilometers)

FIGURE 46 A vehicle-access site on Padre Island, Texas, with numerous tire tracks.
PHOTOGRAPH USED BY PERMISSION OF KATIE PEEK, PROGRAM FOR THE STUDY OF DEVELOPED
SHORELINES, WESTERN CAROLINA UNIVERSITY.

FIGURE 47 The view in the opposite direction from the same vehicle-access site on Padre
Island, Texas, showing the difference in the beach where no driving is allowed. Where driving is
not allowed, the upper beach is vegetated, and small dunes are forming. PHOTOGRAPH USED BY
PERMISSION OF KATIE PEEK, PROGRAM FOR THE STUDY OF DEVELOPED SHORELINES, WESTERN CAROLINA
UNIVERSITY.

apart were built along the coast. The Beach Bill was passed in 1967 and came into being on the basis of long traditional use of the beaches by the state's residents. The bill was inspired by the Texas Open Beaches Act. The Oregon Beach Bill states that "the public may have the free and uninterrupted use thereof" (including for beach driving).

Australia

It is likely that more beaches are open to driving in Australia than anywhere else in the world, although in general beaches with heavy pedestrian use restrict driving. Cable Beach in Broome is a good example of a popular driving beach. This beach has a significant number of rock outcrops which require alert dodging by drivers.

On Fraser Island, police units check speed and occasionally perform breathalyzer tests on beach drivers. The maximum speed is startlingly high (for sand): 50 mph (80 kph). In fact the speed limit was reduced from 62 mph (100 kph) within the past few years!

Instruction pamphlets for drivers on Fraser Island draw attention to the following hazards facing beach drivers:

- Planes landing unexpectedly.
- Small gullies, which are formed by freshwater creeks coming out of the dunes, sometimes changing locations.
- The tides occasionally encompassing the entire beach, leaving no room for driving.

Ireland

Rossnowlagh Beach in Ireland is an interesting case study of the difficulties in overcoming local determination to drive on the beach. In an effort to control vehicles and maintain a Blue Flag award for the beach, the local authority (Donegal County Council) blocked vehicle access to the beach with large concrete bollards. The immediate response from local residents was to bring in heavy machinery and remove the obstacles. Locals objected to their vehicle access onto and across the beach being blocked. Finding no mood for compromise among local residents, the Council undertook a postal survey of almost 7,000 homes in the adjacent area, as well as a survey of beach users. This provided a mandate for controlling cars on the beach and enabled local objections to be overcome in part. A car-free zone was implemented and access across a part of the beach was maintained for local residents.

CONCLUSIONS

Driving is just one of many pressures facing the world's beaches that we discuss in this book. Beach driving has a long tradition, and in many places it is difficult to get to a favored fishing or swimming spot on foot. Even the Surfrider Foundation, arguably the world's most powerful environmental organization dedicated to beaches, makes the case that in many areas access to the beaches for swimming, surfing, fishing or just a day at the beach would be very difficult, if not impossible, without allowing beach driving. Perhaps that is true, but it is also clear that driving on this narrow piece of precious land is damaging. Is this damage worth it? Can't we have a piece of land anywhere without cars and their accompanying noise and pollution and dangers?

It seems clear, however, that as we learn more about the environmental impacts of driving and the fragility of beaches, and as more people use beaches, beach driving is being challenged on a global scale. Steve Ford, a columnist for the Raleigh, North Carolina, *News and Observer*, writing about the driving controversy on the Outer Banks, opined: "Many people want to get closer to nature. And that doesn't mean a beach where vehicles rumble and grind, carving ruts, spewing exhaust and sometimes leaving a nice dollop of crankcase drippings."

Of course, there are beaches and there are beaches. Driving on a crowded Daytona Beach is one thing. An occasional car on a rarely driven-on Moroccan beach is another. Damage is greater on soft beaches, where the uppermost foot (0.3 meters) of sand is churned up, than on hard beaches, where tire penetration is minimal. A special but rare hazard of soft beaches occurred in 2014 on Downhill Strand in Northern Ireland. A car parked on the generally hard, compact beach suddenly sank into the sand, all the way up to the roof. Probably due to water escaping upward through the sand, the beach had almost instantly turned into quicksand and engulfed the car, which, fortunately, was empty of passengers. A similar quicksand event happened at the north end of Kiawah Island, South Carolina. On this occasion, a bicyclist riding on apparently hard beach suddenly found himself up to his neck in sand. He literally had to swim away in the sand to escape.

FIGURE 48 A car parked on the hard sand of Downhill Strand in Northern Ireland suddenly sank into the beach. The beach sand had turned into quicksand. PHOTOGRAPH USED BY PERMISSION OF RACHEL BAIN.

Occurrences such as these may be bizarre, but clearly all beach driving is damaging to the beach. There is an argument that the legal right to beach driving could serve to prevent engineered structures on beaches and might therefore be the lesser of two evils. However, as the case in Texas shows, a legal challenge can immediately change this situation.

As the world's population grows, especially in the coastal zones, more and more people will depend on beaches for recreation, probably for as long as beaches survive. Most of those people will likely be pedestrians, not beach drivers, as the number of drivers is very small compared to those who hang out at a favorite beach spot. Notably, as the sea level rises, beaches will narrow in front of seawalls and disappear, making beach driving impossible.

We are not likely going to be able to preserve developed beaches for future generations, but if we are to preserve beaches for the near future, with their flora and fauna intact, the time has come to reduce and control beach driving. We should follow the example of South Africa of banning vehicles on beaches in an attempt to restore beach ecosystems—and bring back the leopards, at least for a while!

THE ENEMY WITHIN

Beach Pollution

My Name is Ken Seino. I contracted a coxsackie b-4 virus from the Malibu Creek sewage flowing out into the ocean at Surfrider Beach where I was surfing. The creek was spilling out horrible amounts of raw fecal sewage which I couldn't help but ingest as I surfed in the brown and fetid slick covering the surface at third point, Surfrider Beach. I immediately left the water and threw up what nauseous material I could, arriving at the showers to get the pollution off me, my wetsuit, and surfboard. But by the next day I was sick with a temperature of 103 degrees. I stayed sick for three whole weeks, most of that time unable to get out of bed. After 10 weeks of what I thought was a complete recovery, my heart stopped ("second-degree heart block") due to the infection of the coxsackie virus upon the conductive system of my heart. . . . Four other people that I know have succumbed to the very illness that I did [after swimming or surfing on Surfrider Beach]. Two of them are dead now due to the damage this infection does to the right ventricle of the heart. I've been told not to expect a long life because of my infection.
—An excerpt from a 2010 letter written by surfer Ken Seino to the Malibu, California, town board concerning the issue of ridding the city of septic tanks. Seino had a pacemaker installed after his heart stopped.

POLLUTION OF THE SURF-ZONE water and beach sand is a major threat to the long-term future of beaches. More correctly stated, pollution is an increasing threat to the future of long-

FIGURE 49 Raw sewage streaming across a beach in Mumbai, India. There is no worse sight than this on a beach used by swimmers. PHOTOGRAPH BY ANDREW COOPER.

term human use of beaches. Every year thousands of beaches around the world are temporarily closed due to poor water quality. The expected increase in storm intensity in the coming decades, sea-level rise, and added population pressure all point to more beach pollution in the future unless the issue is researched much more fully and new approaches to the problem are developed.

The potential hazards from shark attacks are known to virtually all beach users, and somewhere in the far reaches of people's minds as they step into the ocean is a thought about what large animal might be swimming just beyond the surf zone. An actual shark bite is extremely rare, and a single attack is often international news. But illnesses (some serious) resulting from beach activities are relatively common, yet have certainly not been in the news. It is quite likely that as the growing technical literature concerned with health problems from beach pollution finally makes it to the public eye, the numbers of beach users will drop and their activities on the beach will be of a different sort (e.g., more walking and less getting buried in the sand).

Those who doubt our concern about beach pollution might search the Internet for *beach bacteria* or visit the Beachapedia (an excellent educational resource from the Surfrider Foundation) entry titled "Bacteria in Sand."

BACTERIA: THE GOOD AND THE BAD

According to oceanographer David Karl, there may be a billion bacteria in one cubic inch (16 cubic centimeters) of sand and a million in the same volume of seawater. Most of these tiny creatures are beneficial, helping humans in many ways, including providing oxygen for us to breathe. But along with the good bacteria is a tiny percentage of harmful bacteria, and very nearly all of the bad ones we should fear are of fecal origin. Beach bacteria-related health problems include common gastrointestinal and respiratory distresses, skin rashes, eye ailments such as pink eye, earaches, infected cuts, staph infections (the most deadly of which is MRSA), typhoid, meningitis, Legionnaires' disease, hepatitis, pneumonia, yeast infections, and bloodstream infections. The list goes on and on.

In the last few years, a new dynamic has crept into our understanding of the complex nature of public health risks at beaches. Until recently it has been assumed that the risk is in the water, and we could feel secure (at least to some extent) in the knowledge that the frequent testing of the ocean made our beach trip a healthy one. But now we know that the risk is also in the sand. In a study of three beaches in South Florida, for example, researchers found that fecal bacteria were two to 23 times more common on the wet intertidal beach than in the water column. The concentrations of fecal bacteria on the upper dry beach above normal high tide were 30 to 460 times the concentration in the adjacent surf zone. A few years ago, beach sand was believed to be harmless. In 2006, Chiyome Fukino, Hawaii's state health director, said that there was no point in testing beach sand (after a massive sewage release) because no studies linked illness to bacteria in sand!

The recognition of the role of pollution in sand has turned beach testing on its head. Old assumptions about which beach is the least polluted have fallen by the wayside, and time-tested beach activities, such as digging holes, being buried in the sand, building sand castles, and even the glorious pursuits of lying on the sand and spending the whole day barefoot, must be re-examined.

It used to be swimmer's ear (otitis externa), and then it was swimmer's itch (cercarial dermatitis), or maybe it was coughing and wheezing (respiratory distress). Or a stomachache and diarrhea (gastrointestinal distress) from swallowing too much seawater. (Or was it that bad hot dog?) Anyone who frequently goes to beaches picks up something at one time or another, but now it's a different game. Beach pollution is becoming rampant roughly in proportion to the ever-increasing numbers of coastal dwellers, tourists, and swimmers, as well as increased areas of rain-shedding pavement and rooftops worldwide. New, stricter European Union rules on bathing-water standards have left British authorities with the dilemma of removing the designation of some beaches as "bathing waters" or somehow stopping the agricultural practice of raising spring lambs. (Sheep droppings are the main cause of some rural waters not meeting the new standards.) Meanwhile, a positive element is that new discoveries are leading to a better understanding of the evolution and dispersal of pollutants in the beach system and how to avoid them. Now it is time for scientists to come to the forefront, just as in the case of global climate change, and spread the word.

The annual numbers of reported beach-related illnesses are impressive but are probably low because many people do not report their illnesses or do not make the connection to beaches. Some types of infections may take a few weeks to show up. The Environmental Protection Agency estimates that every year up to 3.5 million people in the United States are sickened on beaches by contact with the vestiges of raw sewage that has overflowed from local sewers. The Natural Resources Defense Council in its annual *Testing the Waters* report notes that during 2011 U.S. beaches were closed because of water pollution for a total of 23,481 days. Some states, such as Florida, announce pollution advisories rather than close beaches. Swimmers are advised of the hazards but are not halted from going into the water.

According to Israeli scientist Hi Shuval, "Every year, bathing in coastal waters polluted with fecal contamination is estimated to cause more than 120 million cases of gastrointestinal illness and 50 million cases of respiratory disease around the world." As many as 1.5 million cases of gastro-

FIGURE 50 The very crowded Haeundae Beach in South Korea. Overcrowding, as in this case, is an obvious source of pollution on any beach. PHOTOGRAPH USED BY PERMISSION OF GERASIMOS NTOKOS.

intestinal illnesses occur every year among swimmers in Southern California alone. A more astounding number is found in the 2012 edition of *Testing the Waters*: "On the basis of beach visitation rates and monitoring data, researchers have estimated that 689,000 to 4,003,000 instances of gastrointestinal illnesses and 693,000 instances of respiratory illnesses occurred each year between 2000 and 2004 at Southern California beaches."

It's not always just stomachaches, as illustrated by the June 9, 2013, adventures of Richard Garey. While fishing for crabs and shrimp in the surf zone of Grand Isle, Louisiana, Garey fell off some rocks and cut his left ankle. He continued to fish but eventually visited an emergency room and got an air cast put over the wound. Forty-eight hours after the hospital visit it became apparent that there was an infection, and it was not an ordinary one. Flesh-eating bacteria (*Vibrio vulnificus*) found in warm salt water were already actively eating his skin and releasing toxins into his body, and Garey came close to death. After seven surgeries and hyperbaric oxygen treatment, he eventually received skin grafts to repair the damage.

Floridian Henry "Butch" Konietzky was not so lucky while crabbing near Ormond Beach. The same type of bacteria that had affected Garey caused an infection that led to Konietzky's death within three days, on September 23, 2013. The bacteria apparently entered his body through a severe bite that he had recently received by stepping on a fire-ant mound. According to the Centers for Disease Control and Prevention (CDC), *Vibrio vulnificus* carries a 50 percent fatality rate. The Florida Department of Health reported that between 2005 and 2009, 138 people in Florida were infected with this bacteria and 31 died. Nine people died from the same bacteria in 2011, as did 13 in 2012. In 2013, 31 people in Florida had been infected with the deadly bacteria by October and 10 had died. These are startling and apparently unpublicized numbers.

The Florida Department of Health says that people with suppressed immune systems (e.g., those with cancer, diabetes, or cirrhosis of the liver) need to be aware of the potential hazards of *Vibrio vulnificus*. In other words, people with suppressed immune systems should not swim, especially if they have an open wound. The widely held urban legend that immersion in salt water helps to heal small cuts and scratches may be true, but is it worth the risk? Patty Konietzky has begun a campaign to publicize the bacterial hazard. Although they were Florida natives, neither she nor her husband had heard of the problem, and now she is pushing the local county commission to post warning signs to inform people about the bacteria.

Fecal Indicator Bacteria

Regulatory agencies check the water quality off beaches by determining the abundance of easy-to-analyze fecal indicator bacteria (FIBS) in the water column. Two common FIBS are *E. coli* (*Escherichia coli*) and ENT (*Enterococcus faecalis*). Both are indicators of pathogens that are more hazardous but more difficult to analyze—that is, the presence of either one indicates that more dangerous microbes, such as viruses or other bacteria, are likely to be present. The abundance of *E. coli* correlates well with bacterial infections in swimmers even if the *E. coli* itself was not responsible for the health problem.

E. coli is a bacteria that inhabits the intestines of humans but also survives well in beach sand, more so in freshwater beaches than in saltwater beaches. *E. coli* generally follows a fecal-to-oral path, meaning that it may be derived from unclean hands. ENT is another bacteria that normally inhabits the intestines of humans and other animals, but in other parts of the body it can cause serious infections.

Measuring FIBS would appear to be a sensible approach to measuring the safety of swimming waters, if sampling is sufficiently frequent, is timed appropriately, and if the analyses are accurate and are reported honestly. But measuring only the water misses the most important source of contamination: the beach sand itself. Only within the last decade have scientists begun to look seriously at how FIBS evolve in beach sand as opposed to the water in the surf zone, and the news isn't good. One 2008 study by Hartz and colleagues shows that both *E. coli* and ENT actually grow and survive better in beach sand than they do in seawater.

The Natural Resources Defense Council urges standardization of water-quality monitoring and beach-closure standards so both tourists and locals can stay informed about the safety of the beaches they're visiting. Enforcement is always a problem on beaches ordered closed by pollution, and sometimes it is lacking altogether. Israeli journalist Roy Arad visited Bat Galim Beach in Haifa and noted that few beachgoers paid attention to the "No Swimming" signs, possibly because the signs were in Hebrew, a language few of the tourists were likely to understand. A lifeguard, when asked by Arad why people swim in the polluted water, responded: "You know how Israelis are; they also drive through red lights." A beach inspector responded to the same question: "I'm in the inspection line, not a lifeguard. I'm not allowed to get involved with the water."

Fecal bacteria in beach sand arrive and leave the sand in a number of ways, as summarized in a 2011 article by scientists Elizabeth Halliday and Rebecca Gast from the Massachusetts Institute of Technology and Woods Hole Oceanographic Institute (MIT/WHOI). They were motivated to start their investigation as a result of the rupture of a sewer pipe on a bluff near the Torrey Pines Golf Course in California, during which three hundred thousand gallons of raw sewage poured onto the beach. Shortly after the spill, a colleague of theirs, Steve Elgar, contracted a serious infection just by walking on the contaminated beach with a small cut on his leg. Clearly, since he had not walked in the water, it was the beach sand that was harboring the bad bacteria that made Elgar ill.

Fecal bacteria get into beach sands from a variety of sources:

- Storm-water runoff carrying bacteria from streets, parking lots, lawns, pet waste, and farms;
- Failing septic systems;
- Boating waste;
- "Bather shedding" by swimmers;
- Local rivers and streams;
- Exchange of bacteria from water into the sand and vice versa;
- Seaward groundwater flowing through the beach;
- Fecal events by dogs, birds, and humans.

Many communities treat sewage and storm-water runoff in the same treatment plant. In the case of heavy rainfall, the plants are overwhelmed and sewage is released with storm-water runoff.

Various methods have been tried to reduce dog waste on beaches. Educational campaigns and provision of pooper scoopers and bins are the main efforts, but some beaches are closed to dogs for part or all of the year. Australia has many beaches specifically designated as dog beaches, and some California beaches, such as Dog Beach at Ocean Beach, San Diego, and Baker Beach near San Francisco, advertise their dog-friendliness. (A genetic study carried out at Dog Beach suggested a link between dog feces and human infection.)

Dogs on beaches are not always popular and can create tensions. In Portstewart, Northern Ireland, a mass protest was held by dog walkers angry at the local council's plans to ban dogs from beaches. In Barbados, concerns over high levels of dog waste on beaches led to a heated online debate between those for and against banning dogs on beaches. Ironically, a study in Wisconsin, where a human-health risk is posed by high levels of seagull waste, an experimental deployment of dogs on the beach as a deterrent to gulls resulted in a marked reduction in bacterial pathogens!

After a long and contentious political battle which began decades ago, Malibu, California, home of the aforementioned surfer Ken Seino, has partially banned septic tanks as part of a water-cleanup operation. Much to the surprise of city officials, a U.S. Geological Survey scientist discovered that birds, not humans, were the main source of fecal bacteria in local waters (except when sewer spills occurred). Pigeons around Hanauma Bay, Oahu, Hawaii, have taken the fall for fecal contamination of the beach there. In La Jolla, California, cormorant feces is a pollution problem. So many birds crowd onto rocks lining the beach that the rocks are white with guano and local shorefront restaurant customers complain of the odor.

Dubai had a special problem. The city expanded so quickly that it completely outstripped the capacity of the single sewage treatment plant. In 2009, sewage was trucked daily from thousands of septic tanks and brought to the central plant. Unfortunately, some drivers had to wait up to three days to discharge their load, so they began dumping raw sewage into storm drains and behind dunes in the desert. Storm drains emptied the pollution onto beaches, prompting beach closures and warnings from local doctors about the dangers of swimming in local waters. Apparently, this situation has been largely cleaned up by the construction of new sewage treatment plants in the area.

Bacteria can also travel. A study of fecal bacteria on Huntington Beach, California, by some researchers shows that FIBs were still viable 3 miles (about 5 kilometers) down the beach from the source, in the direction that the wave-formed currents took them.

A study by UCLA scientists Jennifer Jay and Christine Lee found that some Southern California beach sand had significant levels of fecal bacteria, while offshore waters were "clean." That finding was not surprising, but they also noted that sheltered beaches with low waves, where children like to play, tended to have sand with higher bacteria counts.

A group of scientists from Stony Brook Marine Sciences Research Center studied the abundance and types of yeasts in three south Florida beaches (Hobie, Ft. Lauderdale, and Hollywood) and suggested that the high levels of yeast they observed "may have a deleterious bearing on human health." The most crowded beach had the highest yeast concentration, and on all 17 sampling occasions in 2001 and 2002, the dry beach had a higher density and greater diversity of yeast than the wet intertidal beach.

Just knowing that sand can be polluted should be an eye-opener, particularly for those with suppressed immune systems. However, just because the infectious agents are present in the beach sand or water doesn't mean you are going to be infected. Everyone has a different level of natural defenses against the infectious agents in beach sand or anywhere. This is demonstrated when attenuated (live) vaccines cause a very small number of patients to catch the disease against which the vaccine was supposed to protect.

Sometimes our routine water testing doesn't pick up all the dangers. Recent sampling in the Baltic Sea by scientist Gerald Schernewski of the Baltic Sea Research Institute shows the alarming presence of deadly viruses. These were not on the list of parameters routinely monitored in the water, so authorities had unwittingly been pronouncing the waters safe for bathing simply because they hadn't tested for these viruses. Although viruses are expected in all surface waters affected by sewage input, Schernewski found that polio viruses attached to suspended particles remain infective for weeks and can be transported over long distances.

One end point of the fecal contamination of beaches is found on the beaches of atoll nations in the Pacific, especially around the larger atolls such as Tarawa, Republic of Kiribati (population 50,000), and Majuro, Marshall Islands (population 25,400). Jeffrey Goldberg, in a 2013 *Bloomberg Businessweek* article, notes that in Tarawa he observed (and smelled) piles of human, dog, and pig excrement throughout the crowded village, including on the beach. Apparently, defecating on the beach is popular because "there's a breeze and a nice view and water for washing." A few years back the same phenomenon was observed on Majuro Atoll. The authors have seen human-excrement piles on beaches in many other developing countries, such as India, Colombia, Mozambique, Morocco, and Jordan. This is a widespread problem and is likely to be contributing heavily to the health problems of swimmers, clammers, and strollers in certain countries.

RED-BAG WASTE

Disposal of waste generated in the diagnosis and treatment of humans or animals is a big problem everywhere. In the United States, such wastes are required to be stored in bright red bags before disposal in approved ways. The waste may consist of discarded vaccines, used bandages, cultures and stocks of infectious agents, human tissues, bodily fluids, animal

FIGURE 51 A barefoot girl navigates the garbage-strewn beach in Bajos de Haina, Dominican Republic. The risk of cuts to this child is high, and cuts will increase her susceptibility to bacterial infections. PHOTOGRAPH COURTESY OF EDUARDO MUNOZ / REUTERS.

carcasses, and used and unused sharp objects of various kinds—a dreadful mixture you wouldn't want to see on beaches. Unfortunately, sometimes you do.

Medical waste has been spotted on Imperial Beach in San Diego, California, said to be one of the most polluted surf spots in the United States, due to runoff from sewage emanating from the Tijuana River. In 2009, local surfers were urged to get vaccinated against hepatitis A. Lake Michigan communities were warned of a medical-waste spill from overflowing sewers in Milwaukee, Wisconsin, where heavy rains often result in sewage spills. Of course, dangerous syringes are often thrown away by drug users.

At least twice in recent years, needles, syringes, used bandages, and a variety of other medical throwaways appeared on some New Jersey beaches, including those of Sandy Hook and Sea Bright. The assumption was that the material was illegally dumped at sea by waste barges. In 1987, a 50-mile-long (80 kilometers) household-garbage slick was found off the coast of New Jersey, and a variety of medical wastes floated from it onto the nearby beaches. Since that time, medical waste disposal has been more carefully monitored. Yet, because so many industries and nations consider the sea to be a low-cost landfill, this awful stuff does sometimes end up on beaches, regulations or no regulations.

MRSA

MRSA is the acronym for methicillin-resistant *Staphylococcus aureus*. This form of staph infection is the scourge of hospitals, as it is difficult to keep from spreading and difficult to treat in infected patients or in those who have been exposed to it. Now it has also been found on recreational beaches, probably left on sand by other bathers. In the past it was assumed to be restricted to warm subtropical waters, but now it is known in the cold, cold waters as far north as Pacific Canada.

MRSA has become very personal for Orrin Pilkey, whose 26-year-old grandson, an avid surfer, picked it up from Westport Beach in the state of Washington. He cut his foot on an underwater object but continued surfing. Upon arrival back home, he put antibacterial medicine on the wound and figured it would clear up. The leg began to swell and was very sore, so he soaked it in hot water and continued to go to work every day. Running a fever, one day he collapsed at work. His boss took him to the emergency room, where the wound split open and the emergency room was closed and declared contaminated. He had contracted MRSA, a word he had never heard before.

He has recovered and still surfs but now must take medication and will have MRSA the rest of his life, with flare-ups expected from time to time—a stiff price to pay for a few hours of surfing in cold waters where no one knew MRSA even existed.

At an American Society for Microbiology meeting in 2009, Marilyn Roberts, a University of Washington microbiologist, reported that at least five public beaches along the coast of Washington were contaminated with MRSA, found both on the sand and in the water. Because of the uniformity of the MRSA type on several beaches, she suspected that some of the bacteria may have been derived from non-human sources.

Dr. Lisa Plano and associates, from the University of Miami's Miller School of Medicine, reported in 2013 that 37 percent of water samples from swimming beaches and 25 percent of beach-sand samples in their study contained *S. aureus*, including some MRSA. More MRSA was found in the dry beach-sand samples than in the intertidal zone. The specific beach in South Florida was not identified, although a number of photos were featured with the article. The lack of clear identification of the studied beach may be a reflection of the political dynamite of identifying MRSA in a beach from a state so economically dependent on beaches.

Not to worry, said several California newspapers, no MRSA found on our beaches. Well, not exactly. In comments posted to the September 12, 2009, *Los Angeles Times* story on MRSA ("Is Your Beach Contaminated with MRSA?"), Glenn Rock said:

> I live in La Jolla and swim almost every day in the ocean at the La Jolla Shores Beach. About a week ago my arm/shoulder were so badly infected that I went to the emergency room. I hadn't hurt myself and it just came out of the blue. It turns out that I had a Staph (antibiotic-resistant strain) infection. The lab at Scripps called me several days after taking some tests to tell me that my infection was very serious because of the resistance to antibiotics. Well now after 2000 mg (4 horse pills) of Keflex a day and 2 huge pills of Bactrim, tons of creams, etc. prescribed my infection has gotten much better and hopefully will be all gone in the next 3 or 4 days.

He did not identify his staph infection as MRSA.

TOXIC ALGAE

A relatively unknown menace that is mounting an assault on beaches is large accumulations of green algae, which are a reflection of water pollution. Humans are behind this one, too. According to various media reports in France and the United Kingdom, lethal green algae have invaded heavily used vacation beaches in Brittany (northern France) and along the British coastline from Wales to Portsmouth. Lying in deep piles up to 3 feet (1 meter) thick with hard crusting on top, these stinking seaweed masses are ticking gas bombs.

Vincent Petit, a 27-year-old veterinarian, was riding horseback on a Brittany beach near Saint-Michel-en-Grève when his horse broke through the crust atop a mass of algae and went down. A cloud of hydrogen sulfide gas was released from the rotting algae, reportedly killing the horse within 30 seconds. Fortunately, a tractor that was used to clear away the algae was nearby and dragged Petit to safety. He was rescued in an un-

FIGURE 52 Green-algae accumulation on a Brittany, France, beach resulting from overfertilized farm fields. As the algae decompose, they produce hydrogen sulfide, which has killed animals and sickened people. PHOTOGRAPH USED BY PERMISSION OF CRISTINA BARROCA.

conscious state and hospitalized. Now he is suing the local municipality responsible for beach maintenance.

The lethal algae on the French coast were apparently a product of the over-fertilization of nearby fields with drainage emptying directly into the ocean. Towns along the Brittany coastline have hired bulldozers to scrape the seaweed away, but the algae keep coming back. Earlier, on a beach close to where Petit's horse died, two dogs strolling by were killed by the hydrogen sulfide. In strange coincidences indicating the global nature of this problem, the deaths of two dogs running on an algae-encrusted beach north of Auckland, New Zealand, were recently reported, and four dogs were killed in 2009 by toxic beach algae near Elkton, Oregon.

The more one learns about this beach hazard, the more apparent its global scope becomes. In 2008, the Chinese government brought in 10,000 workers to clear away the slimy green growths (at a cost of US$30 million) so the Olympic sailing competition could be held and observers could safely view the event. In 2013, an even larger algal bloom occurred on the beaches around Qingdao in Shandong province along the Yellow

Sea. Beachgoers seemed to enjoy walking and even rolling around in the green slime, but local officials were cleaning it up as quickly as they could. By late June they had removed close to 20,000 tons of algae. It turns out that the nutrients introduced to the sea from local farms may not have been the source of the bloom. Algae farms to the south grow a specific type of alga valued by Japanese diners. In the farming process, unwanted algae species are removed and disposed of in the Yellow Sea, where they thrive and rapidly spread.

In Italy, near Genoa, a 60-year-old man had to be taken to the hospital because he swam in algae-infested water, and in Genoa, more than 200 people were sent to the hospital after swimming in the algae or inhaling toxins carried to the beach by the wind. During the summer of 2012, officials in Massachusetts put out a toxic beach-algae warning but did not close the beaches. This is a problem for freshwater lakes as well.

Some are attributing the algae outbreaks to global warming. Although this may indeed be a factor as our seas warm up, it is clear that excess nitrate-rich fertilizers, along with animal wastes and poorly treated or untreated sewage, are the main villains. Hot weather, warm water, and fertilizer runoff from farms near rivers that run into the sea are the problems, but these tend to disappear with the arrival of fall and winter. Unfortunately, it is always summer or spring somewhere on our planet, and the problem flows from the land to the sea. More than 70 beaches in northern France are in trouble, as is the British coastline in agricultural areas. In Galveston, Texas, where, in season, tons of seaweed are removed from beaches nightly, efforts are being made to compact the seaweed and use it to construct dunes.

The problem is deeper than just hazards to humans. When a beach is covered with algae, virtually everything that lives on and within the beach is killed, while access is denied to nesting and food for local birds, fish, sea turtles, and various crustaceans. Thus, an entire beach and nearshore ecosystem, from microscopic organisms (meiofauna) living between sand grains at the bottom of the food chain up to sharks cruising offshore, is either wiped out or forced to relocate. Simultaneously, oxygen is usually depleted in nearshore waters, a further threat to marine mammals and seabirds. In China's Shandong province, farms of abalone, sea cucumbers, and clams have been devastated by algal blooms.

THE RED TIDE

Red tides are a type of algal bloom of phytoplankton in nearshore waters. When the concentrations of the poisonous plankton are large, the water may turn red. This global phenomenon is caused by a population explosion of dinoflagellate plankton that produces toxins. Red tides occur in the warm-water season and are favored by calm weather, low salinity due to recent rains, and high nutrient loading from farming. It is a natural process, but we include brief mention of it in this chapter because global climate change and nutrient pollution, which impact the red tide, are both human-related.

Red tides are hazardous to humans because they cause eye and respiratory irritation, as well as coughing, sneezing, and itching, in people standing on beaches. In our experience, the first sign of a red tide near a beach is involuntary coughing. Once when standing on a North Carolina beach with a group of students, we all simultaneously realized that most of us were coughing lightly, a strange experience. We had been forewarned about a red tide and we fled the beach immediately.

The toxins involved in red tides can cause massive fish kills and have also been known to kill manatees, sea turtles, and dolphins. Birds feeding on dead fish washed up on the beach are also victims of the red tide. Red-tide toxins can accumulate in oysters, mussels, and clams, sickening those who eat them and leading to the closure of clam and oyster fisheries.

Red tides can be devastating to the local tourist industry. The 2005 red tide in southwest Florida caused tourists to avoid the sights and smells of the beaches covered with dead fish. Emergency room visits increased, and those who ventured near the beach suffered immediate raspy throats. The effects of this red tide lasted for more than a year.

One of the most deadly toxins in red tides is saxitoxin, which can cause paralysis of the lungs of humans, obviously a fatal problem. In 1990, six fishermen on Georges Bank off New England saved the mussels they had inadvertently caught in their fishing trawls and ate them in a grand meal at the end of a long working day. The captain wandered into the galley a bit late and immediately observed that his crew members were entirely incapacitated and all were having trouble breathing. He may initially have suspected that they were drunk, since the early symptoms of red-tide illness mimic drunkenness. The captain, who had also consumed some mussels, managed to call the U.S. Coast Guard, which carried the men to the nearest hospital. All were saved, but it was a near-death experience from a red tide believed to have originated in shallow water.

HOOKWORMS

Cutaneous larva migrans (CLM), otherwise known as creeping nematodes, may be the most widespread infection that humans can pick up on beaches. This is a type of hookworm (*Ancylostomatidae*) that crawls under the outer layers of skin and continues to creep around until it dies weeks or months later. It goes by several other strange names, such as creeping eruption, ground itch, and plumber's itch. The infection results in bright-red and intensely itchy skin patches, but uncomfortable as it may be, it is not fatal. The nematodes commonly affect the feet or buttocks, body locations most likely to be in contact with beach sand.

Even the best beaches are not immune to hookworms. In October of 2010, South Beach in Miami suffered an outbreak believed to have been caused largely by cat feces in dune sands.

Millions of people around the world are probably affected every year by hookworms caused by walking barefoot on beaches. It is a warm-water problem, mostly on beaches in Southeast Asia, northern Australia, Central and South America, the Caribbean, and the southern United States. The worms arrive on the beach in the feces of dogs and, to a lesser extent, cats. (When is the last time you saw a cat on the beach?) The solution is to wear shoes and sit on (multiple) beach towels or on beach chairs and to make sure that pets are dewormed.

HOW TO USE A BEACH

In this brief survey of pollution on beaches, we have only scratched the surface of a subject worthy of a book in itself. We've glossed over several important aspects of pollution, such as the problem of stagnant water that is created updrift of coastal-engineering structures (e.g., jetties and groins, and also in groin fields). And lurking in the shadows are sea-level rise and the flooding of landfills and hazardous-waste sites located at low elevations, which will have a potentially huge impact on all aspects of beach quality and safety.

In writing this chapter, both of us were shocked by the extent of pollution on beaches far from big population centers and by the known ex-

FIGURE 53 Nuclear power plant in Dungeness, United Kingdom. Nuclear plants require water for cooling, and as a consequence a number of them around the world are next to ocean beaches. One nuclear power plant is located on a barrier island on the east coast of Florida. As has been shown by the 2011 tsunami disaster in Japan, such plants are a major potential source of pollution. PHOTOGRAPH USED BY PERMISSION OF JOSEPH T. KELLEY.

tent of illnesses caused from beach visits. On some beaches, fecal bacteria from birds (e.g., the pigeons in Hanauma Bay and the cormorants in La Jolla) and animals can be as deadly as storm-water runoff from urban beaches. One study looked at the number of beach-related maladies by age groups and found that fewer than 10 percent of beach dwellers are affected in every age group, and that the vast majority of afflictions are minor. But at the same time, a significant number of illnesses are major, as in the case of Pilkey's grandson's MRSA infection, Garey's tangle with flesh-eating bacteria, and Seino's heart problem. All were products of beach pollution. Surfers Against Sewage, a U.K. organization, identified more than 900 cases of significant illnesses caused by contamination of U.K. beaches from 1998 to 1999.

The recent discovery that sand can hold more fecal bacteria than surf-zone waters points to the need for an entirely new approach to beach monitoring. One wonders why the troubling aspects of the science of beach pollution have not been publicized. As mentioned earlier, one study showing MRSA in Florida's water and sand did not identify the beach. It goes without saying that the beach-tourist industry might not be excited about such revelations. With rising coastal populations, beach illnesses will continue to increase in proportion to numbers of swimmers and abundance of the bad bacteria. Perhaps it is time for scientists to speak out—even if no one asks them to!

More studies are needed on the distribution of fecal bacteria and other pollutants. Although virtually every study of which we are aware has found relatively high concentrations of dangerous bacteria in beach sand, a wider geographic spread of samples is needed. Can we remediate a beach that is widely used yet highly contaminated? Is there a machine similar to the beach-raking tractors found on major tourist beaches that could clean the sand (but not at the expense of the ecosystem)?

How can we safely have fun on the beach? The goal should be to avoid contact, as much as possible, with fecal bacteria known to lurk in the sand. This being the case, common sense might argue:

- Don't go barefoot on the beach (but this is probably unthinkable for most beach lovers).
- Never lie or sit directly on the beach sand (perhaps this is equally unthinkable?). Instead, use multiple beach towels (hookworms can burrow through a single towel), or sit on a beach chair.
- Never allow yourself to be buried in beach sand. This may be the worst practice of all, as virtually your entire body is in contact with the sand. Building sand castles may be a problem as well, especially for children.
- Those with cuts and open sores should avoid beaches altogether.
- Dogs on beaches should be free of worms, particularly hookworms.

University of Miami researcher Lora Fleming has a different list of dos and don'ts on the beach—rules to stay healthy:

- Avoid getting seawater in your mouth.
- Don't swim when you are ill.

- Shower before and after your visit.
- Wash your hands before eating.
- Take small children to the bathroom frequently.

The Louisiana Department of Health and Hospitals has published a number of swimming tips, including:

- Don't swim near drainage pipes, cross-beach runoff streams, or littered areas.
- Avoid swimming after heavy rains.
- Don't swim with an open cut, wound, or skin infection.

The U.S. Geological Survey, in a study of stomachaches related to beach visits, says that the simple procedure of thoroughly washing one's hands before eating would reduce gastrointestinal stress. The Centers for Disease Control and Prevention (CDC), as well as officials from the Florida Department of Health, warn that flesh-eating bacteria are particularly dangerous to people with suppressed immune systems, especially if they have an open wound.

Research your beach. In the United States, check the annual Natural Resources Defense Council's *Testing the Waters* report, and in Europe check the Blue Flag system. And finally, avoid very crowded beaches.

THE FUTURE OF BEACH POLLUTION

Beach pollution will increase in coming decades and will become a major cause of long-term beach closures, and in some cases, abandonment. Increasingly, beaches will be closed in response to higher values of FIBs in the water. When officialdom finally recognizes the high FIB values in the dry sand beach, there will certainly be an increase in beach closures. The upside of this is that if a lot more closures occur, there will be public pressure to do something about it, to determine the exact sources of pollution, and to devise measures to reduce the problem.

In part, the cause of increased pollution is escalating world population, but sea-level rise will play a major role as well. As the sea rises, pollution will increase while shorelines retreat over various waste-disposal sites and the sites of former beachfront buildings that caved in or were demolished. Ever-increasing areas of impervious surfaces of paved roads and roofs will provide larger amounts of storm-water runoff flowing to the beaches, often mixed with raw sewage from overwhelmed sewage-treatment plants. Expected intensification of hurricanes combined with more people living near the coast will cause the release of ever-larger amounts of raw sewage, as experienced with Hurricane Sandy along the East Coast of the United States in 2012. Such fiascos, along with increased use of fertilizer in fields near the beaches, may increase toxic accumulations of algae.

Trash on the beach is in itself a form of pollution. Its ever-increasing appearance on the world's beaches is a reflection both of our growing population and its affluence. The great tsunamis of the last decade have produced a vast volume of trash that is slowly wending its way to the world's beaches or to the great garbage patches in the central areas of the world's oceans.

9 THE INTERNATIONAL DIMENSION OF BEACH DESTRUCTION

WHEN IT COMES TO CONTEMPLATING the international aspect of beach destruction, images of warfare often come to mind. Many famous military actions took place on beaches, for example, beach invasions in Gallipoli, Normandy, and Iwo Jima, as well as the evacuation from the beach at Dunkirk. The bombing, detonation of mines, and passage of thousands of tanks and other vehicles caused devastation that is well-illustrated in countless movies. As damaging as those actions were for the people involved, from the perspective of the beach, the effects were largely ephemeral. Each of those beaches is still there, and the damage inflicted by a few weeks of warfare was long ago repaired by natural beach-recovery processes. As it turns out, the international dimension of beach destruction is much more pernicious and much longer-lasting than mere warfare.

FIGURE 54 Four decades ago, Benidorm, Spain, shown here, was a sleepy fishing village. Its character rapidly changed to accommodate the many German and British tourists who discovered the warmer climate. Now the community is similar to a high-rise-lined shoreline of Florida, with the looming problem of responding to a higher sea level. Will the buildings be moved, demolished, or simply abandoned? PHOTOGRAPH USED BY PERMISSION OF ARNOLD BROWN.

TOURISM

International tourism is dominated to a great extent by the lure of beautiful beaches, and yet its impact on those beaches has been devastating.

Since the British first "discovered the continent" and began to take European vacations after the Battle of Waterloo (1815), they have left a major impact on mainland European beaches. Of course much of this was undertaken from the best of motives. The early British holidaymakers at Nice on the French Riviera first constructed a path on the top of the shingle beach. This was subsequently replaced by a seawall and promenade, the Promenade des Anglais. These were built in the early years of the 1820s and were financed by a British clergyman (Rev. Lewis Way) who

FIGURE 55 An aerial view of La Manga del Mar Menor, Spain, an overcrowded strip of land that was practically undeveloped a few decades ago. Many of Spain's barrier islands and spits, all low-elevation bodies of sand, are densely developed, taking advantage of a tourist industry based on Europeans from colder climes. Much of the Spanish development on the Mediterranean is highly susceptible to sea-level rise. PHOTOGRAPH COURTESY OF GREENPEACE ESPAÑA; USED IN ACCORDANCE WITH GREENPEACE'S VALUES.

was anxious to find work for the many impoverished people in the town. The subsequent loss of the supply of gravel from the rivers to the beach (because of dams on inflowing rivers) resulted in its rapid erosion and narrowing, which then threatened the promenade. Since 1976 the beach has been artificially maintained by adding more than 720,000 cubic yards (550,000 cubic meters) of gravel, a consequence that was certainly not foreseen by the early tourists.

Unfortunately, from these early beginnings beach tourism has extended around the world with similarly damaging impacts, all because of the refusal to take account of the mobile nature of shorelines. Beach tourism has evolved from initial small-scale development of hotels on the dunes and beach to high-rise development on the shoreline. For example, there is now an 80-story beachfront condo at Surfers Paradise on Australia's Gold Coast, with a second 80-story building planned. Alongside intensive beachfront construction projects, tourist beaches are now being shaped according to human needs rather than nature's, and along with the accompanying misconceived shoreline-defense works, every bad practice is being exported in this era of globalization.

As discussed in chapter 2, on small islands where the supply of construction material is limited, there has been a long tradition of taking sand from beaches for construction purposes. In many cases, particularly in the Caribbean, along both coasts of Africa, and throughout the tropics, this might once have been sustainable, since the beach sand there is often made up of the fragments of local mollusk shells and coral. Thus the sand is constantly being resupplied from the birth, growth, and death of seashells in the warm tropical ocean. International tourism, however, prompted a dramatic increase in the need for construction sand on tropical islands, in many cases outstripping the rates of fresh-sand production. As a consequence, many beaches on islands, such as Puerto Rico, Saint Croix, and Barbados, have become narrower or disappeared altogether. In some places the beach-damaging effects of sand mining can take years to materialize, but in other places they are almost immediate.

We have observed beach sand being removed by dump truck immediately in front of a plush tourist resort in Ecuador. In Kenya, we watched sand being taken from the beach and used to make concrete for a seawall for an adjacent beachfront hotel. The effect was to narrow the beach and make the wall (and the hotel) more vulnerable to waves than before. This kind of misguided effort (robbing Peter to pay Paul) that tries to save tourism-related infrastructure misses the fact that removing sand from the beach in front of the hotel destroys what tourists have come for *and* weakens the beach's natural resilience to storms.

In 2008, the BBC reported a remarkable difference in how local people and foreign holiday homeowners in Zanzibar responded to natural beach retreat. Villagers abandoned houses closest to the sea and rebuilt at the landward side of the village, an apparently traditional practice that is quite sustainable. Inspired by the idyllic tropical location, some European tourists had built a few holiday cottages next to the native village. When these inevitably became threatened by erosion, rather than retreat as the villagers had done, the tourists built seawalls in an effort to protect their investments. In so doing, they exacerbated the erosion problem for themselves and their neighbors and destroyed the beach in the bargain.

FIGURE 56 An artificial urban beach in Dubai, lined by a jetty and a rock revetment.
PHOTOGRAPH BY ORRIN PILKEY, FROM PILKEY, ET AL., *THE WORLD'S BEACHES*.

This is a small-scale example of the lengths to which some will go to protect even small-scale tourism infrastructure. Much of global beach tourism involves wealthy and powerful countries exploiting countries that are less so. When it comes to bigger developments, the impacts are even greater. We have dealt with the specific negative effects of seawalls and beach replenishment elsewhere in this book, but here we are principally concerned with the international dimension of tourism. One of the major influences globally is the construction of tourist developments on foreign beaches by international companies. Beach tourism is a major international business—just consider the distribution of the large hotel chains around the world's beaches (e.g., Hilton, Holiday Inn, Ramada, Sheraton, Le Méridien, Iberostar, Wyndham, and Westin). While beach tourism in the past often grew organically to some extent, there is now a much more formal and dedicated approach to its development. In some cases this involves entire cities (such as Dubai) or islands (such as those in the Maldives), but often beach tourism development involves only a part of the coast or a single beach. We might expect that a dedicated and specific approach such as this could avoid the mistakes that arise as beach resorts develop organically. Far from it.

FIGURE 57 A Santa Pola beach near Alicante, Spain, which has been raked for cleaning and aesthetics. Such beach cleaning is a disaster for the beach ecosystem. PHOTOGRAPH USED BY PERMISSION OF NORMA LONGO.

The architects who design major tourist resorts operate on a global scale. They appear to have their own ideas of what will work at beach resorts, irrespective of the natural processes that have shaped and sustained those beaches in the past. In the architects' efforts to fit in the desired hotel rooms, swimming pools, and other infrastructure, the natural beach is often ignored. Instead, the beach seems to be treated as a garden with pleasant curves inserted to maximize the attractiveness. And usually there is no government agency or environmental group to object.

Architects and politicians seem to believe that beaches must be engineered, so groins, walls, and breakwaters of all sorts are added to create the idyllic beach resort of the architects' and developers' dreams. Daily beach cleaning, often by rakes dragged by tractors, assures the complete destruction of the beach ecosystem.

Consequently, many island beaches that are widely perceived as idyllic are actually very poor imitations of the real thing. For example, few of the islands in the Maldives that have tourist development are free of rubble seawalls, dredged channels, groins, and reclaimed land because of the desire to maximize the size of the development.

International beach tourism is plagued by corruption, as wealthy developers get involved with local politicians and officials of developing countries with the power or influence to enable developments to proceed. In this scenario, the chances of preventing measures that destroy the beach are almost zero. Beach-tourism development has often been easy to "sell" to people in impoverished coastal areas. The promise of jobs and income often blinds the citizenry to the more damaging aspects, and the impacts can often take years to become evident.

Another damaging aspect of beach tourism is the demand for souvenirs. Often this involves exploitation of the local supply of shells, coral, and other materials. When these run out or are subject to restrictions, coral and shells are imported from countries where they are not protected or where protection is poorly enforced. The foot-long (0.3 meters) Bahamian conch (queen conch) is found in almost every shell shop in the Americas and Europe. Cebu in the Philippines is a center of seashell collection and export for the tourist trade, with dozens of souvenir shops and exporters based there, including U.S.-based shell dealers. In a series of raids in early 2011, authorities targeted illegal exporters of seashells in Cebu and recovered nine sacks of *Tridacna* shells (the giant clams), three boxes of *Murex* shells, three sacks of sponges, a box of coral, and four baby sharks. *Tridacna* shells are protected under the Convention on International Trade in Endangered Species of Wild Flora and Fauna (CITES). There have been several high-profile prosecutions in the United States of people involved in the illegal importation of tons of seashells.

That is not to say that beach tourism is necessarily always damaging. Low-impact beach tourism exists; however, this tends to be the exception rather than the rule. There are cases where the influence of foreign visitors has been important in preventing damage. International surfing groups, such as Surfriders, and various conservation organizations have successfully campaigned to preserve some beaches from poorly planned development.

BEACH SAND: THE INTERNATIONAL TRADE

A small secluded beach known as Cleopatra's Beach on Sedir Island off southwestern Turkey contains sand that is quite unlike that of all other nearby beaches. Tradition has it that the sand was brought from Egypt as a gift for Cleopatra by Mark Antony. This might be the earliest case of taking sand from one country to another to build a beach. True or not, the story draws tourists to the beach—in fact it draws them in such numbers that the Turkish government has placed entry restrictions, and visitors must rinse their feet in basins as they leave to ensure that none of the precious sand leaves the beach.

Perhaps the ancient Egyptians initiated the trade, but whether they did or not, the international trade in beach sand is now big business. In the past, for example, beach sand has been imported to Waikiki Beach from Los Angeles, Australia, and adjacent Hawaiian islands. The extent of beach replenishment around the world means that sand is now far more in demand and more valuable than when it was simply used in construction. Consequently, an international trade in beach sand has sprung up in recent decades, which is discussed in chapter 4. Usually this involves the export from poor countries to their wealthier neighbors who have run out of sand for beach replenishment. Often poorer countries have weak or poorly enforced environmental legislation to protect their own beaches. The huge demand for sand in Singapore has led to numerous claims of illegal practices and corruption in the acquisition of sand from Vietnam, Malaysia, Indonesia, Burma, the Philippines, Bangladesh, and Cambodia (see chapter 2). The demand for sand has been blamed for environmental degradation in several of these countries. In Cambodia, dredgers have been reported inside nature-conservation zones (the Peam Krasaop Wildlife Sanctuary and the Koh Kapik Ramsar Site), and damaging impacts on mangroves, sea-grass beds, and coral reefs have all been reported. A decline in fish stocks in Cambodia has also been linked to the practice of sand extraction for export.

On the other side of the world, the tiny Caribbean island of Barbuda is selling the family silver, in the form of beach sand, to other Caribbean islands (see chapter 2). As Barbuda has tightened its laws regarding the export of sand, Guyana is now being seen by some as the next supplier of much-needed beach sand. One company (Guyana Sand) advertises "very white and clean" beach sand for sale, among its other products.

The Canary Islands, Spain, are one of Europe's top beach-vacation destinations, with 12 million tourists visiting each year. Following a now-standard model throughout the Canary Islands, a series of beach resorts mostly with replenished beaches have been built on the south coast of Gran Canaria, at Amadores, Anfi del Mar, and Puerto Rico. In each case an unremarkable stretch of rocky coast was converted to a resort by building a marina and a beach. The beaches are made of imported sand, from either the Caribbean or Morocco, and each is held in place by two massive groins that converge at their seaward end. These developments are owned by international companies that follow the same approach around the globe.

Sometimes nature is behind the international sand trade. An unanticipated two-way international trade in replenished beach sand has been established between the German Frisian Island of Sylt and the adjacent island of Rømø in Denmark (see chapter 4). Longshore drift transports the replenished German sand north, to Denmark. Ironically, the Danish beach was already regarded as among the widest in Europe.

In a more fragile setting, most of Israel's sandy beaches are produced by the transport of sand from the Nile Delta in Egypt. The damming of the Nile and the reduction of sediment supply to the delta are thus having an impact within and far beyond the borders of Egypt. Similarly, in Portugal beaches are denied fresh supplies of river sand because many of them have been dammed upstream in neighboring Spain. Beaches on the Mekong Delta in Vietnam are expected to erode extensively because of sand-trapping dams already constructed in China and soon to be constructed in Laos.

The mining of beach and dune sand for heavy minerals by international companies in the developing world often exploits the lack of protection afforded to the beach environment. All mining does not create long-term problems for beaches. Gold mining in Nome, Alaska, and diamond mining in Namibia obviously do not remove a large volume of sand. The volume of gold and diamonds in these sands is minute, hence the beach should recover and return to its natural processes a few years after the miners have left.

Given the accumulated experience in Europe (beginning with the Romans) with coastal engineering and the construction of sea defenses, it is perhaps not surprising that these practices that have been so damaging to the continent's beaches are being exported. Engineers are falling over each other in their bids to sell engineered "solutions" to beach erosion around the world. Almost without exception, the engineering approach to solving shoreline problems involves maintaining the status quo—that is, holding the shoreline in place with little concern for the coming sea-level rise. In addition, the engineering approach favors buildings over beaches.

On Australia's Gold Coast, a major storm in 1972 caused a great deal of damage to beachfront property. Who did the local council turn to but Delft Hydraulics (now called Deltares) in the Netherlands to come up with a protection scheme? Unsurprisingly, the scheme involved the construction of a dike to protect property and beach replenishment to sustain the beach. The Delft Report is still the mainstay of Gold Coast beach management. But being the most experienced coastal engineers in the world does not make the approach of the Dutch a good solution to every erosion problem. Far from it—in the Netherlands, sea defense is a matter of national survival, and its current situation has evolved from centuries of interaction with the sea. Seawalls, dikes, and replenished beaches that may work well for the wealthy Dutch society may not work at all in other political and economic realms.

After Hurricane Katrina wreaked havoc in Louisiana, advice was sought from the ever-willing Dutch engineers. The same thing happened again after Superstorm Sandy in New Jersey and New York in 2012. All of Europe's big engineering companies are vying for business worldwide, and most have subsidiary offices in all parts of the world. Deltares engineers even point out the particular suitability of several of their "alternative approaches" for the developing world. DHI (its Danish competitor) proudly reports on their beach work in Brazil, the Bahamas, India, Oman, Iran, and the United Arab Emirates.

In May of 2012, another European operator, Euroconsult Mott Mac-Donald, reported its role as project manager for a US$45 million project on coastal protection in Karnataka, India, where about half of the 186-

FIGURE 58 A church in Tamandaré, Brazil. Originally built back from the sea, the church is now virtually on the sea's edge and is in serious trouble. This is a common scene throughout older coastal towns and cities in Brazil. PHOTOGRAPH BY ANDREW COOPER.

mile (300-kilometer) coastline is retreating. Despite talk of long-term sustainable measures in the project press release, the approach involves holding the shoreline still with shoreline armoring, beach replenishment, and offshore breakwaters.

In 2008, the World Bank funded a US$5 million project to stop coastal erosion in Vilanculos, Mozambique, thus preventing the sustainable option of retreat from the shoreline and initiating the eventual demise of a beautiful sandy beach. In Maputo (Mozambique's capital city), a US$21 million project funded by the Arab Bank for Economic Development in Africa and the Saudi Development Fund will build eight breakwaters and 6 miles (10 kilometers) of seawalls and repair an existing seawall around a beach. The message seems to be, if you don't have the funds to mess up your own beach, there are others with the finances and know-how to do it for you. The message is also: No need to worry about the rising sea level now. We'll think about that later.

Perhaps the biggest example of this trade is evident at Dubai's Palm Islands and World developments, where millions of tons of sand were dredged from the seafloor to create artificial islands, protected by massive boulder revetments and seawalls for development of tourism. Thousands of apartments, villas, and hotels are being built on these artificial islands and peninsulas. At its peak, some 40 percent of the world's dredgers were working on the developments in Dubai. Aside from the damage done to the seabed and the creatures that once lived there, the artificial islands, the peninsulas, and the plethora of groins, breakwaters, and seawalls along mainland Dubai have totally destroyed the natural beaches along their part of the Persian Gulf Coast. Not only that, but the costs of protecting all of this against future sea-level rise will be astronomical, even if it were technically feasible. It is somewhat ironic that at the same time that the Maldivian government is seriously contemplating the possibility of moving to a new homeland in the face of rising sea levels, and while efforts to preserve Venice from rising sea levels are fraught with problems, new problems are being actively created in Dubai.

INVASIVE SPECIES

A commonly unappreciated international influence and problem on beaches is the spread around the world of exotic species that then outcompete native species and become invasive. European settlers in the western United States, South Africa, Australia, and New Zealand were perturbed by the bare sand in the coastal dunes. To stabilize the dunes they imported European beachgrass (marram grass—*Ammophila arenaria*), which was more vigorous in its growth than the native species. The grass did the trick and stabilized the dunes, but it also changed their shape, forming much higher dunes than had previously been present. The new foredunes cut off the sand supply to the dunes behind the foredunes, causing them also to stabilize with vegetation and altered the entire sedimentary system. In addition, the diversity of the native vegetation was markedly reduced because the wind and salt-spray factors in the ecosystem changed.

European beachgrass was imported from northern Europe around 1869 to stabilize dunes in San Francisco's Golden Gate Park. For more than a century, this beach grass colonized most dunes on the Pacific coast of the United States through widespread planting enhanced by the plants' robust natural spreading capability. *Ammophila* collects sand more effectively than the native grass, *Leymus mollis* (American dune grass), and thus steep and tall continuous foredunes, up to 33 feet (10 meters) high, formed along the coast. Just like in Australia and New Zealand, these large foredunes cause sand starvation of active inland dunes. The rapid spread of *Ammophila* on the dunes themselves has reduced the area of open, active unvegetated dunes on the coast.

In South Africa and the Caribbean, the approach to stabilizing bare sand and establishing forests on dunes was to import *Casuarina* trees (Australian pines) that grew very well around the Indo-Pacific. They thrived in the new areas, but their unexpected side effects soon were manifest, one being that a thick covering of needles blanketed the underlying sand, killing the native vegetation. Not only that, but the trees advanced across the beach, right up to the normal high-tide line, reducing the dry-beach width and leaving little recreational area or space for turtles and birds to nest. It has also been speculated that shade from the trees might alter the sex ratios of turtles emerging from nests since the ratio is temperature-dependent. Furthermore, during hurricanes *Casuarina* trees are readily toppled, sometimes blocking key evacuation routes, and, particularly in Florida, the trees are regarded as invasive.

CLIMATE CHANGE AND DEVELOPMENT AID

With the recognition of the reality of global climate change (at least among the "believers") and its potential impacts in the developing world, there is much concern about helping developing nations cope. The argument is often stated that developing countries didn't create the problem, but they have to live with it. It is a worrisome but sad fact that this is leading to more and more sea defenses being constructed with development-aid money. "Sea defenses" often destroy the natural resources on which developing countries depend.

On a small scale, on the almost-undeveloped island of Anegada in the British Virgin Islands, a badly sited beachfront development of six rental bungalows was partly destroyed by shoreline retreat. Futile efforts to save the structures using sand-filled bags (geotubes) were funded using money allocated by the British government for climate-change adaptation.

Sadly, there are countless examples of this flawed thinking, both on the part of the governments seeking international aid and those granting it. The European Union allocated €55.4 million for sea defenses in Guyana from 2008 to 2013; Britain allocated £60 million for sea defenses and farmland protection to Bangladesh in 2009; the government of Switzerland (a landlocked nation) plans to contribute US$3.25 million to "protect" the coastline of the city of Beira in Mozambique; and US$2.9 million were granted to the island of Kiribati by the United Nations for shoreline protection under the heading of "increasing coastal resilience." Climate-change adaptation seems to be regarded as a license to build sea defenses to protect property without regard for the impacts on beaches, or for the sustainability of the approach. Of course, if someone else is paying, the long-term economic considerations are forgotten, in much the same way that early development-assistance schemes resulted in lines of broken-down tractors, abandoned because there was no money to repair them.

THE FUTURE OF BEACHES

The future of beaches around the world is a direct function of the type of development. For example, many popular tourist beaches in warm waters are in developing countries. The Maldives and the Caribbean islands already are stabilizing their shorelines with little thought to future sea-level rise. Their future beach-conservation efforts are constrained by lack of money and a job-hungry population.

The selling of beach-mined sands across international boundaries is a major blow for beaches in some developing countries. Such beach mining not only destroys beaches, it also provides a strong incentive for corruption in local governments.

Perhaps the most fundamental global problem is the prevalent attitude that buildings are more important than beaches. Around the world, wealthy and influential people cluster along the shorelines. In order to save the buildings, creative and well-trained coastal engineers are called upon, particularly from the Netherlands and Denmark. As an old saying goes, if you call upon a surgeon, you are likely to have surgery. The same goes for beaches. If you call upon engineers, you are likely to get an engineering "solution," and the shoreline will be held in place. Holding a shoreline in place means, on a generational scale, giving up on the beach.

Most important of all is the role of international aid organizations

FIGURE 59 Is this the future of beaches? The person in the foreground has found solitude while sunbathing next to the sea on a rock revetment in Monopoli, Italy. PHOTOGRAPH USED BY PERMISSION OF NORMA LONGO.

concerned with helping developing countries adapt to climate change. It is vital that these organizations, plus donor and recipient governments, understand that holding shorelines in place during rising sea levels is the worst thing to do. The best help will be in finding alternative forms of adaptation so that developing countries do not repeat the mistakes of the developed world. Already in several developing countries (e.g., Nigeria and the Pacific coast of Colombia), beachfront homes are designed to be easily moved back away from the retreating beach. This is a locally devised solution, not one devised by engineers, and it is the only type of solution that will allow development and beaches to coexist.

THE END IS HERE | 10

THERE ARE TWO POSSIBLE scenarios for the future of the world's beaches. One is based on the philosophical view that buildings are more important than beaches, a view that has provided the basis for most of the world's beach management over the past century. As we have outlined in this book, the world's beaches face many problems—including erosion, hard stabilization, replenishment, beach degradation, and pollution—and most of these problems are related to coastal development.

The other scenario holds that beaches are more important than buildings and that buildings must make way for beaches. The absence of beachfront development means fewer problems, as well as more beautiful beaches for our great-grandchildren to enjoy. But our beaches have been so heavily controlled, lined with large buildings, shaped, mined, armored, and buried in dredged sand for so long that it is

FIGURE 60 Moving the Schifter house in Chappaquiddick, Massachusetts, in 2013. While the move was taking place (it cost more than the original construction), the shoreline was held in place by the sandbag revetment on the beach. The house was 220 feet away from the bluff edge when it was built in 2007. PHOTOGRAPH USED BY PERMISSION OF BILL MCGONAGLE, CAPE COD AERIAL PHOTOGRAPHY.

easy to lose sight of the fact that they once were a wonderful and complex natural environment much like a rainforest or a coral reef.

We believe that the current outlook, biased toward protection of property, will inevitably lead to a worldwide loss of beaches lined with development. We might have gotten away with this outlook on beaches for a few more decades, but the warming oceans and the melting glaciers and ice sheets have changed all that. The sea level is rising, and our beachfront buildings are threatened as never before. It now comes down to buildings or beaches: we must make our choice.

THE FUTURE OF BEACH QUALITY: A SUMMARY

The quality of beaches comes down to factors that are partially independent of beach management but that are ultimately linked to development. These include all the issues discussed in the foregoing chapters: trash on

FIGURE 61 Moving the "Serendipity" house that was threatened by shoreline erosion in Rodanthe, North Carolina. This house was featured in the movie *Nights in Rodanthe*, during which a storm occurred and the actors stayed in the house, a foolish storm response for anyone in a beachfront house. PHOTOGRAPH USED BY PERMISSION OF JIM TRIBBLE.

beaches, oil spills by tankers and drilling rigs, sand and water pollution, and driving on beaches. Trash is most likely to increase in proportion to the population. Natural disasters such as the Indian Ocean tsunami of 2004 and the Japanese tsunami of 2011 produce surges in the quantity of beach trash. The worst is yet to come from the Japanese debris field still drifting relentlessly across the North Pacific. Meanwhile, the "great sea" of floating plastic in the central oceans (the plastisphere) will continue to grow and occasionally spin off trash to beaches.

Trash creates more than an unsightly mess. The Japanese trash coming ashore in Alaska is often rich in small Styrofoam particles, which are ingested by both fish and marine mammals; and food containers with long-spoiled contents add to the trash problem. Even if it can't be controlled at the source, trash can easily be picked up and disposed of, but on tourist beaches trash removal is often done by raking, sometimes with large mechanical rakes dragged by tractors. This is a fatal process for the beach fauna that live within and between the sand grains, and it cuts off the bottom of the food chain for the nearshore marine environment.

The cleanup of beach trash (including ubiquitous cigarette butts) is best done the old-fashioned way—picking it up by hand so that the seaweed can be left to power the beach ecosystem. It is also important to occasionally clean up remote beaches affected by particular events, such as those beaches in Alaska covered by debris from the Japanese tsunami, or the seemingly remote beaches of islands such as Molokai, Hawaii, which regularly accumulate floating debris from across the South Pacific.

While quantities of trash increase on beaches, oil on beaches, often in the form of tar balls, has undoubtedly diminished over the last 50 years or so due to better shipping and drilling regulations. Nonetheless, there will continue to be major bumps in the road, such as the BP oil spill of 2010 in the Gulf of Mexico. Also, there are some rogue freighters and tankers that still dump oil on a regular basis. Tar balls will always be with us on beaches, but probably not in increasing numbers in coming decades.

In the future, pollution of beaches, both the water and sand, will increase as coastal populations grow. Trash, oil, and mining are things that can be cleaned up or controlled, but the pollution of water and beach sand is another matter. And the extent of the problem remains unknown to the general public.

The storm-water runoff in fast-growing beach communities, which one way or another makes it to the beach, will increase, and expected larger storms will create ever more runoff and release more pollutants than ever. Many beaches are already unswimmable, or at least should be. A recent example of this problem occurred with Hurricane Sandy, which, according to Climate Central, released about 10 billion gallons of untreated and partially treated sewage into the waters around New York and New Jersey.

In 2014, as we were finishing writing this book, the British Marine Conservation Society issued a startling report on beach pollution. Of 754 U.K. beaches tested, only 413 were recommended for swimming. The Society said that the fall in water quality (113 fewer beaches were recommended

FIGURE 62 A careful look at this photograph shows four water pipes sticking out of the sea, each marking the site of a house that has fallen in on (appropriately named) Washaway Beach, Washington. At least two blocks of houses in this community have been swept away by the ocean. PHOTOGRAPH USED BY PERMISSION OF NORMA LONGO.

for swimming compared to the previous year) was due to a very wet summer. The resultant flooding caused an increase in pollutants "from a variety of sources such as agricultural and urban runoff, storm waters, plumbing misconnections, septic tanks and dog waste."

We anticipate that there will be increased recognition of the pollution hazards to beach-goers as the problem becomes more obvious and as the already-large technical literature on the subject finally makes it to the public eye. With increased recognition must come increased monitoring of both beach sand and surf-zone water. Currently, only water (not sand) is routinely tested for fecal bacteria, and even the water isn't routinely tested for other potentially harmful organisms and viruses.

The bottom line is that there will likely be increased pollution of the world's recreational ocean beaches in coming decades. A change in behavior on beaches will be evident as public recognition of the problem increases. Unthinkable as they may seem today, the rules will likely be:

1. Don't walk barefoot on the beach.
2. Don't lie on the sand. Lie on a thick towel or blanket.
3. Never, ever get buried in beach sand.

Another dilemma, beach driving, creates problems for turtle- and bird-nesting, can be damaging to beach intergranular fauna, and interferes with pedestrian use. The number of beach users who navigate on foot are orders of magnitude larger than those who drive. In the eyes of some, beach driving is an activity that is in itself an eyesore in this unique natural environment. Driving on beaches probably will remain a popular pastime, but with coastal populations increasing and ever-rising beach popularity, driving will likely be squeezed down to smaller areas.

As to beach-quality changes, the principal problem is that when the beach is lost, so too is a large and important ecosystem, together with all the benefits it brings to the human population.

THE FUTURE OF THE BEACH ITSELF

There are four important rules to follow in order to preserve beaches for future generations in a time of rising sea levels:

1. *Do not build seawalls*: The most significant cause of the complete disappearance of beaches has been and will continue to be seawalls. Seawalls built on eroding shorelines always destroy the beach.
2. *Do not build beachfront high-rises*: Multistory condo and apartment buildings anywhere on barrier islands or within locally determined setback distances on other beaches completely reduce any flexibility in response to sea-level rise. High-rises, for all practical economic purposes, cannot be moved, and usually there is no place to move them. High-rises lead to seawalls.
3. *Do not mine sand*: Preventing sand mining will become increasingly difficult in the future. Cheap, easy-to-mine sand is a juicy target for the construction industry, and mining the sand destroys beaches.
4. *Value the beach ecosystem*: The impact of seawalls and especially beach replenishment on the beach ecosystem is barely considered. Not only does beach replenishment kill the entire system but if the new beach sand has a different grain size from the original, a different ecosystem will arrive.

FIGURE 63 The writing is on the wall. *Nothing is permanent*: an appropriate saying on a seawall on a beach in Mumbai, India. PHOTOGRAPH BY ANDREW COOPER.

A few decades ago, the standard "solution" to the erosion threat to buildings on ocean shorelines was to build a hard structure, such as a seawall or groin. Seawalls always destroy sandy beaches, as we learned from experiences in Europe and New Jersey, and groins are not much better over the long run. Sadly, however, both are still seen as the primary solutions to beach erosion in many places in the world.

For instance, on French Polynesian atolls, piles of coconut-palm logs face the open ocean. Wide wooden planks are braced on their sides to prevent spring tides and minor storms from flooding poverty-stricken villages on Colombia's tropical barrier islands in the Pacific. Lines of abandoned cars and trucks on a beach on Majuro, Marshall Islands, attempt to hold the line, and household items, including kitchen sinks and abandoned dogsleds, line some beaches in indigenous villages along Arctic shorelines.

FIGURE 64 These post–Superstorm Sandy 20- to 25-ton boulders are part of a future seawall in Southampton, New York—a community that has tried to outlaw seawalls. When completed, this revetment, intended to protect a single home, may well be the mightiest coastal-engineering structure on the East Coast of the United States. It is an example of how wealthy homeowners can override local governments. PHOTOGRAPH USED BY PERMISSION OF ROB YOUNG, PROGRAM FOR THE STUDY OF DEVELOPED SHORELINES, WESTERN CAROLINA UNIVERSITY.

As the sea level rises and larger numbers of buildings are threatened, seawalls are making an inevitable comeback in the developed world. We expect that 30 to 40 years from now, seawall construction and repair will be a major global industry. In the next 50 to 70 years, miles upon miles of beaches will have disappeared, along with their nearshore ecosystems. Birds will find fewer beaches to feed on, and birds and turtles will find fewer nesting places, except possibly in parks and nature preserves.

Very often the impetus behind seawall construction, even when it will clearly lead to beach degradation, is furnished by the wealthier elements of a community. Such is the case in Southampton, New York, on the south shore of Long Island, a community that has long opposed seawalls. Taking advantage of a post–Superstorm Sandy beach-regulation loophole, at least five Southampton homeowners were building massive seawalls in 2013 to protect their homes, which are valued between $25 million and $65 million. One homeowner, the hedge-fund billionaire Chris

Shumway, is using boulders, each weighing between 20 and 25 tons, to build what may be the mightiest seawall of all on the U.S. East and Gulf Coast shorelines—all this to protect a single house.

Inevitably, one seawall always leads to more seawalls, so Southampton is on its way to a beachless line of seawalls. Once you start you can't stop is the rule. Fred Havemeyer, a member of Southampton's board of trustees, is quoted in the *New York Times* as saying: "If you lose the use of the beach, you've lost Southampton. All these people are extremely rich and they're broadcasting the message of 'Me first.'"

Solana Beach, California, is a community that recognizes the threat of seawalls to the community's future. The town has passed an ordinance requiring a time limit for approval of future seawalls or modifications of existing seawalls that protect bluff-top development. The structures will be approved for only 20 years, after which a new permit will be required. In other words, seawalls are not to be viewed as permanent fixtures. This of course exacerbates uncertainty in the real-estate business. Removing the walls will be required if they are causing damage to the beach (a foregone conclusion). Property owners are suing the town, noting (correctly) that the new requirement greatly lowers beachfront-property values.

Another example of the role of wealth in driving the beach-management trajectory is Figure Eight Island, North Carolina. Figure Eight is the most exclusive island in the state. Through the residents' influence and via contributions to the campaigns of local politicians, the state's anti-hard-stabilization rules were changed to allow jetties (renamed *terminal groins*) at the ends of Figure Eight and other islands. It is clear that the jetties will only protect the first dozen houses nearest the structures on Figure Eight and will increase erosion rates elsewhere.

After Hurricane Sandy slammed into New Jersey, New York, and other parts of the Eastern Seaboard, there was considerable discussion about the need to recognize the long-term impact of sea-level rise and not simply rebuild on the same footprint. In fact, this discussion was more prominent than in any aftermath of an American hurricane. The more normal response to such storms is the complete rebuilding of the community as it once was. Unfortunately, it has already become clear that seawalls will be an important element in the post-hurricane response. They range from walls in front of individual houses to a 13-mile (21-kilometer) wall along the central coast, to be built by the U.S. Army Corps of Engineers. Many

FIGURE 65 The Outlaw family house in Nags Head, North Carolina, is said to have been moved back from the ocean five times, for a total of 600 feet (about 180 meters). In the early moving operations, the house was rolled along on logs while being pulled by mules. PHOTOGRAPH BY ORRIN PILKEY.

of the proposed walls will be covered with sand after construction to reduce the visual impact. Experience shows, however, that the sand covering is often removed in the next storm. A seawall is still a seawall whether covered by sand or not.

One Connecticut newspaper suggested that after Superstorm Sandy, seawalls were desperately needed to protect nature—a clear misunderstanding of the fact that walls are built to protect houses, not nature. Choosing the seawall alternative is a sad and retrogressive response to storm damage and wholly inappropriate at a time of rising sea level. Far from protecting nature, they confound it.

Another response to the loss of beaches is the soft "solution"—beach replenishment—which introduces new sand to widen the beach. Such artificial beaches usually erode at least twice as fast as natural beaches. After a time span that varies from place to place (but is always fewer than 10 years and is usually closer to three years), the beach will have to be replenished again. As time goes on and as the sea level rises, the interval of re-replenishment will get shorter because the beach becomes less stable,

Table 1. Future Beach Replenishment Volume Estimates for Gold Coast Beaches
for Different Sea-level Rise Scenarios

Sea-level rise	1 meter	2 meter
Initial sand (cubic meters)	18 million	36 million
Additional (cubic meters)	180,000	360,000
Annual cost	$10.8 million	$21.6 million

A rough estimate of the costs and sand volumes of an 29-kilometer-long beach (18 miles) in Gold Coast, Australia, under three sea-level-rise scenarios. Additional sand refers to the annual amount required over and above existing volumes. Costs assume A$60 per 1.0 cubic meter.

and at the same time, offshore sand supplies will diminish as they become harder to find. Compromise sand of unsuitable grain size will begin to find its way to beaches. This is already happening in North Carolina, where both too muddy and too shelly beaches have been pumped up. Sea-level rise assures that the end of replenishment will soon arrive, except for the few beach communities that can afford high-cost sand. The annual volumes of replenishment and especially the costs have increased steadily in the Netherlands as they seek to hold the shoreline in place. A possible beach-replenishment cost scenario for the future of beach replenishment on the Gold Coast of Australia beach is shown in table 1.

BEACH TIME LINES FOR THE FUTURE

There are expected time lines for the future condition of different types of beaches. The assumption is that the sea level will have risen around 3 feet (1 meter) by the year 2100, and so we consider the scenarios for 30, 60, and 90 years into the future for developed, high-rise-lined, and third-world beaches carrying out business as usual—where we continue to treat beaches in the same way as at present. After this outline is an alternative scenario based on a change in attitude that is expressed as retreat from the shoreline and an entirely new view of beaches.

Business as Usual on Developed Shorelines

If we continue to carry on in the same way as we do now, defending buildings and other structures, beaches will likely evolve in the following way.

THIRTY YEARS FROM NOW

Most shorelines have seawalls, including those where seawalls were previously forbidden. Most beaches are gone at high tide except for the few communities that can still afford to replenish their beaches, but even these are now in an unstable position due to sea-level rise. Frequent re-replenishment means that the beach ecosystem cannot recover, and there are major reductions in fish and shellfish stocks. Saving the cities has become a higher priority than saving beach tourism communities, limiting federal funding for the latter.

SIXTY YEARS FROM NOW

Seawalls have become mighty fortresses behind which beachfront properties shelter. Their views of the sea are obscured by the height of the defenses. Seawalls extend completely around islands. Beaches are strewn with debris from fallen seawalls, like the New Jersey beaches of the 1960s. Very few beaches are still being replenished—sand is expensive and in short supply. Even beaches in nature preserves have seawalls. The primary beach activity is promenading on top of seawalls, enjoying the sea breeze and salt spray. Turtles and bird populations are in trouble. Beach-dependent fish and shellfish (e.g., New England steamers and Dover sole) are extinct. Much pollution is present, from increased storm water and urban and agricultural runoff to organic enrichment of coastal waters due to the loss of beach-filtering mechanisms.

NINETY YEARS FROM NOW

Beaches have largely been reduced to rubble from decaying seawalls. Swimming conditions are very dangerous. Old men tell stories of their boyhood fun on the beach to incredulous grandchildren. Significant amounts of money are being spent on maintaining seawalls, and "beachfront" communities face financial ruin. The best beach tourism is restricted to far-away and remote places, such as Namibia, Chile, and Siberia. Funding from a central government for shoreline engineering has disappeared completely except for cities.

Business as Usual on High-Rise-Lined Shorelines

The constraints imposed by high-rise development limit management options still further. The density and value of property in high-rise developments create huge economic arguments for defense, even in the face of sea-level rise.

THIRTY YEARS FROM NOW

This is the worst-possible situation for preserving beaches for future generations, best exemplified by the coasts of Spain and Florida and increasingly by other U.S. Gulf states. These shorelines are being held in place as long as economics permit, but the increasing cost and environmental degradation will eventually lead to their demise. Large seawalls have been built on the oceanfront, and small seawalls or levees have been built on the bay sides of communities. All remaining beaches are replenished—no natural beaches remain, and beach ecosystems are devastated. Communities are struggling financially, as the costs of maintenance and replenishment increase and the environmental quality declines. The lower stories of high-rises have no sea view.

SIXTY YEARS FROM NOW

Massive seawalls surround entire beachfront communities. A huge pollution problem develops as water that seeps under the walls is pumped out. Few beaches remain, and those that do are almost devoid of life. Property values decline in response to the degraded environment, and maintenance costs increase. The desirability of oceanfront living declines.

NINETY YEARS FROM NOW

No beaches remain. Seawall rubble litters the shoreline. Walls around communities make them resemble prisons. People must climb to the top of the seawall to get a view of the ocean. Vertical seawalls are a particular high-tide hazard for children and pets. Swimming in front of the seawalls is impossible because waves are breaking strongly on the walls and the rubble. The ocean water is heavily polluted. Blocks of high-rises are falling into disrepair and being abandoned.

Business as Usual on Third-World Beaches

Beaches in the developing world face many of the same issues as those in the developed world, but important decisions regarding their future state will be influenced by the direction of development-aid funding, particularly funding related to climate change. This in turn will be strongly influenced by politics.

There are two scenarios here—one where international aid pours into a country in a vain and misguided effort to hold the shoreline in place, and one where aid is instead used to encourage retreat. A deluge of international development aid could spell the end for beaches in the developing world, where hitherto a lack of finance has been beneficial in ensuring that seawalls are not built and that communities have moved landward with rising sea levels. It is likely that, given the choice, property owners will opt to have their homes defended, especially if someone else is paying. This will be encouraged by engineering companies looking for business. In this scenario some beaches may survive, but they will be littered with debris from ineffective seawall attempts. Extensive mining of the beaches is likely, as sand supplies for construction are used up. Some sand will even be sold to replenish beaches in developed countries where sand supplies have run out. Pollution in the groundwater, the beaches, and the ocean will be high as various waste sites are flooded.

If development-aid funding is used to assist with retreat, the outcome will be beneficial—beaches will survive by moving landward, buildings will be demolished or relocated, and the beaches will form the basis of a new global tourism market, as traditional beach resorts elsewhere are bereft of beaches. Fishing will improve and local economies boosted.

Business as Usual on Remote, Undeveloped Beaches

THIRTY, SIXTY, AND NINETY YEARS FROM NOW

Beaches will adjust to changing sea levels and remain in perfect natural shape—providing that miners in search of sand for construction and replenishment are kept out.

The Retreat Option

If we acknowledge that protecting buildings is harmful to beaches and that buildings have to make way for beaches, the outcome for all types of beaches and for those who use them or benefit from them will be dramatically different.

In response to shoreline retreat, beachfront buildings are moved or demolished. Since local communities almost always prefer beach loss to building loss, the retreat option must be governed and enforced at a higher level of government. Initially, this is a politically difficult process. Prohibiting reconstruction of beachfront buildings destroyed by hurricanes or storms is one approach, as is prohibiting hard defenses for any new building threatened by erosion. Buildings and roads are threatened. Much depends on how well seawall and beach-bulldozing restrictions are enforced. The presence of high-rises in a particular location effectively prevents this alternative response (at least until costs of defense become too high). Clear responsibility must be established as to who will quickly clean up buildings that collapse. Beaches will survive in variable conditions according to how effectively infrastructure that is threatened is removed.

Erosion threatens coastal landfills, which pose a pollution threat to the beach if not carefully removed.

SIXTY YEARS FROM NOW

Building demolition has become more common than moving buildings since relocation sites are now more rare. Beach pollution from demolition activity, flooding of toxic waste sites, and discharge of storm water runoff all have to be managed. The costs, however, are much less than continued defense, and beaches survive intact.

NINETY YEARS FROM NOW

By now, because major coastal cities are critically endangered, funding is difficult to find to relocate buildings along recreational beaches. Beaches on sandy coasts are well inland of their locations ninety years earlier. On steep, rocky coasts, there is much less beach retreat. In either case, the ecosystem is largely intact and thriving, and the beach is also surviving. Swimming and other recreational opportunities persist.

THERE IS A RIGHT WAY

In the effort to preserve beaches for future generations, the United Kingdom's National Trust is one organization leading the way. Clearly, this requires sacrifice and hardships that do not sit well with some coastal dwellers. The National Trust (a non-government equivalent of combined Nature Conservancy and the U.S. National Park Service) is adapting to sea-level rise; land is being lost and structures, even historic buildings, are being sacrificed. And local citizens are going along with it, a great accomplishment in itself. The National Trust is leading the way in coastal adaptation. In the publication *Shifting Shores* they state, "There is no guarantee that hard defences work in the long term: they are often only a temporary solution. As sea levels rise and severe storms increase, it will become ever more difficult and expensive to build and maintain strong defences. They can also disfigure the coast and cause environmental harm by moving the problem to another location. We believe therefore that hard defences should only be used as a last resort."

In 2008 the tabloid newspaper the *Daily Mail* reported that the National Trust was abandoning a nudist beach to the sea. Beach huts had been moved back three times in response to shoreline erosion, and there was nowhere left for them to go on the narrow strip of sand that is Studland Beach in Dorset. Years of experimentation with sea defenses of various kinds had not only been in vain but had damaged the coast. Typical objections from beachgoers (e.g., "I don't understand how they are going to just let the beach be taken away by the sea.") underlined the public's lack of appreciation of the link between beaches and the sea. Inspired by rapid development and degradation of the coast, the National Trust began a campaign (Operation Neptune) in 1965 to buy areas of the British coast for the purpose of protecting and conserving them for future generations. As a result, the National Trust currently owns 742 miles (1,194 kilometers) of coastline in the United Kingdom (about 10 percent of the coastline of England, Wales, and Northern Ireland). Faced with the combined issues of flooding and erosion as the climate changes and sea levels rise, the National Trust is assessing how best to respond at its sites. Their 2006 policy of non-intervention in natural processes, however, recognizes the inevitability of coastal change and that preserving beaches means allowing them to interact with the waves and giving them the space to change shape and location as needed.

The change in policy considered the long-term aspects of management action at the shoreline and advocated making important policy decisions as early as possible to minimize environmental damage and reduce costs:

"There are many reasons why adapting [a new policy] as early as possible is desirable. It is likely to be the most realistic and cost-effective approach over the long term. It helps people understand the risks they face and gives them time to adjust and adapt their communities, so reducing the risk of suffering catastrophic flooding and erosion."

The steps that have been taken in the U.K. show a major departure from past practice, and implementing them has required much work with local communities and government at different levels. As well as preserving natural processes in its own properties, the National Trust's approach is showing others what can be achieved and how it might be done. The Trust's approaches include changing the design of essential buildings and infrastructure, ceasing bad management practices that resist nature, allowing the shoreline to recede at the cost of buildings, and, most radical of all, removing seawalls and groins. The following is a brief description of how the National Trust maintains the natural environment.

Adapting Buildings

Many of the beaches and stretches of coast owned by the National Trust include buildings and infrastructure (e.g., houses, parking lots, beach huts, lighthouses, roads, and public toilets) whose future needs to be reconsidered in light of shoreline recession.

At Brancaster in Norfolk, an activity center owned by the National Trust is periodically flooded by high tides. The frequency of flooding will increase with the rising sea level. To cope with this, the National Trust decided to live with nature as best they could. The structure was altered so that electrical cables are now in the ceiling, and all electricity sockets have been raised three feet above the floor. The floors were covered with washable materials to prolong the building's life before it eventually has to be demolished.

At Portstewart Strand, Northern Ireland, a new visitor center was built at the beach entrance. It is prefabricated so that it can easily be dismantled and moved should circumstances require it.

Moving Infrastructure at Risk

The beaches of the Studland Peninsula in Dorset attract more than a million visitors each year. The beach is receding by 6 to 10 feet (2 to 3 meters) per year at an area where cafés, toilets, a shop, beach huts, and a parking lot are present behind the beach. The beach huts have been moved landward several times, and the other buildings are being relocated to allow the beach to move and survive.

At Formby in Lancashire, erosion of the dunes by 10 to 13 feet (3 to 4 meters) per year is accompanied by the occasional storm that can cause more than 33 feet (10 meters) of erosion. To allow the dunes to move back naturally and still maintain public access, a footpath and parking lot are being relocated landward.

Allowing Space for Coastal Processes

At Birling Gap on England's famous chalk cliffs, erosion provides a steady supply of flint nodules that feed the adjacent beach. Historic Coastguard cottages were threatened by erosion, and there was pressure for seawalls to be constructed to halt cliff recession. These would have destroyed the beach and degraded the cliff. After several legal battles and a prolonged negotiation, cliff recession was enabled to proceed, at the cost of some of the Coastguard cottages. In 2002 the National Trust demolished a Coastguard cottage on the cliff edge. A small hotel and three more cottages are close to the cliff and will eventually be removed as the cliff recedes.

Restoring Natural Processes and Undoing Past Mistakes

A massive gravel beach at Porlock in Somerset, U.K., used to be maintained artificially by pushing gravel up the beach to increase its height and protect a freshwater wetland behind it. This created an artificially high berm and resisted the natural processes of barrier breaching. A new policy of non-intervention allowed waves to break through the barrier, creating a new inlet and transforming the freshwater wetland into a brackish-water wetland. A similar situation of resisting natural processes existed on a large beach at Man Sands near Brixham. In 1985 gabions were emplaced in a futile effort to halt erosion. In 2004 the gabions were removed and the beach was allowed to re-establish a natural profile. Drainage pipes under the sand barrier were also removed at the same time, allowing a wetland to be re-established behind the barrier.

Abereiddy is a sand-and-gravel beach on the Pembrokeshire coast of Wales. A parking lot on the back-beach had been protected by a seawall since the 1960s, but by 2000 the wall had fallen into bad repair. After consultation with local people and adjacent landowners (who had concerns over flood risk and erosion), the wall was removed to enable natural processes to reshape the beach.

At Mullion Cove in Cornwall, a substantial stone-built harbor constructed in the 1890s has suffered repeated storm damage. Following consultation with local residents and armed with various proposals (including an offshore breakwater), the decision was made to allow the harbor to be gradually removed by not repairing it after future storm damage. Eventually as the harbor is removed, the beach will be restored to its pre-harbor condition.

At Brownsea Island in Poole Harbour (a large estuary), successive attempts to halt erosion in the 1970s used a variety of steel and wooden pilings and gabions. By 2008 the National Trust decided not to replace these but instead removed them to restore natural shoreline processes.

Similarly, at South Milton Sands in Devon, wooden sea defenses constructed in 1990 at the rear of the beach were removed to reinstate the natural connection between beach and dunes. The degraded dunes were replanted with marram grass.

Preserving Ecosystems

The policy of allowing natural processes to continue extends to sustaining beach ecosystems, so in 2007 the practice of cleaning Studland beach by removing sea grass was discontinued. As a result, strandline plants (e.g., sea rocket) have become established and are helping to stabilize the beach and dunes. In response to objections about the "unsightly" and "smelly" seaweed, a community-engagement program was initiated to demonstrate the importance of seaweed for the beach ecosystem.

On the rapidly eroding Suffolk coastline near Dunwich, the National Trust owns a narrow eroding barrier beach that partly protects a bird reserve in a freshwater wetland. The Trust decided to allow natural processes to continue despite the probability that the barrier will break down, causing the wetland to partly lose its natural protection and to change its character.

All of these changes were driven by a desire to work with natural processes, recognizing that this is the way that beaches will survive. The adjustments that people have to make in order to allow this to happen are generally quite minor, and yet these changes were bitterly resisted at the start. By engaging with coastal residents and explaining the long-term benefits of natural processes, the National Trust has been able to implement a fundamental change in approach to preserving the beaches under its control.

Perhaps the U.K. National Trust leads the way because they have an intuition concerning shoreline processes based on such a long history of interaction with shoreline change. In a startling 1912 book, *The Lost Towns of the Yorkshire Coast*, historian Thomas Shepard described 28 small towns, dating back 2,000 years to the Roman invasion, that now reside on the continental shelf. Some towns are now 3 miles (about 5 kilometers) seaward from today's shoreline, under the waves. With a view of 2,000 years of development history, the British should have a good sense of who wins in the end on the shoreline!

RAYS OF HOPE?

Unfortunately, the National Trust way is not the global way. The future of beaches around the world is a direct function of the type of development. For example, many popular tourist beaches in warm waters are in developing countries. In the Maldives and the Caribbean islands, beaches are being stabilized with little thought to the future rising seas. The beach-conservation efforts are constrained by lack of money, unscrupulous developers, corrupt officials, and a job-hungry population.

The selling of beach-mined sands across international boundaries is a major blow for beaches in some developing countries (e.g., Barbuda and Morocco). Such beach-sand mining not only destroys beaches but also provides a strong incentive for corruption in local governments.

Perhaps the most fundamental global problem is the prevalent attitude that buildings are more important than beaches. All around the world, wealthy and "important" people live at the beach. To save their property they rely on engineers to hold the shoreline in place. Losing the beach and depriving other people of its pleasures is of no concern to beachfront-property owners.

The role of international aid organizations is crucial if the beaches of the developing world are to survive. For that to happen we need an enlightened approach that acknowledges that the developed world has made a mess of things and should not export the same misguided approaches.

There are a few glints of light that give cause for hope. In most countries some beaches are protected in national parks or by conservation charities. In the United Kingdom, the biggest coastal landowner is the Na-

FIGURE 66 Is this the wave of the future? This Red Sea beach is gone and has been replaced by attractive stone steps, but the rocks at the base of the steps probably prevent swimming. This is Zeytuna Beach in El Gouna, Egypt.

tional Trust, which has a commitment not to interfere with natural processes. In the United States, the national seashore system of the National Park Service could also lead the way. In 1972 the National Park Service declared that no shoreline engineering would be allowed at the seashores. It was a stunning announcement that met with much complaining when it first came out. In some U.S. states (e.g., North Carolina and Maine), hard defenses are not permitted and beaches are given priority over buildings, but there is a constant battle to retain those enlightened laws when property is threatened. Some of the battles have been lost.

Australian federal policy now requires that beach amenity be regarded as more important than buildings when making decisions at the coast. In a similar vein, an organization called World Surfing Reserves is following the UNESCO World Heritage Program model in establishing a number of world surfing reserves that may afford protection to some beaches. These are all steps in the right direction, but they are needed everywhere, not just in a few locations. Furthermore, the efforts of the National Trust and the National Park Service are generally directed only to undeveloped or lightly developed beaches.

Recognizing the damage that it does, several countries have banned sand mining from beaches and rivers, although in the developing world, these laws are often not enforced. With more stringent control of sand mining, alternatives are being developed, such as machines that can produce sand from quarried rocks. Many initiatives are trying to clean up beaches—from local cleanups to regulation of pollution nationally and internationally—yet beaches continue to be degraded.

Despite these rays of hope, we've only scratched the surface. It is indisputable that as a society we value beaches, but in the battle of buildings versus beaches, buildings continue to win. It is our hope that this book will go some way toward recognizing that predicament. Beaches are reaching the stage of critical endangerment, and in whole regions natural beaches are now long extinct. Major changes are needed if beaches are not to be consigned to the dustbin of history.

A NEW VIEW OF BEACHES

If we continue to treat beaches in the way that we do now, it may not be long before we face the prospect of visiting the last beach—the last in a county, state, country, or even the planet. All would have been extinguished in our efforts to protect the more-ephemeral structures we build behind them. The average house may have a life span of, say, 100 years, a road, somewhat longer, but it is common to see high-rise buildings demolished after 50 years or fewer. The life span of a natural beach, in con-

FIGURE 67 The Morris Island lighthouse near Charleston, South Carolina, was originally built behind the island. Today it is 1,300 feet (400 meters) in front of the island. The island migrated after losing its sand supply, which was cut off by the jetties at the entrance to Charleston Harbor. PHOTOGRAPH USED BY PERMISSION OF MARY EDNA FRASER.

trast, is measured in many thousands of years. And as we have seen, the natural beach provides a range of services to us at no cost. No matter what we install in place of a beach, whether a seawall or replenished beach, it is less efficient and more costly than what it replaces.

We wholeheartedly believe that it is time to look at beaches in an entirely new way. In our new view of beaches, the primary goal for society should be preserving the beach's integrity, with all of its functions and processes. Never again should beaches become engineering projects intended to preserve the accoutrements of humans at the price of beach quality. In this new view, we value beaches more than buildings. We value public beaches over private property. And we put beachgoers (both living and yet unborn) ahead of beach-property owners. We should only use hard defenses as a very last resort around critical infrastructure—and even then only while the infrastructure is being moved.

And if hard structures are used, they should be viewed as emergency stopgaps. We would be ashamed to have to replenish a beach or stabilize an inlet. Public awareness would be raised to the point that the mere mention of a seawall would bring people onto the streets in protest, enraged at the prospect of destroying a beach and its ecosystem.

We must view the beach as a sacred and resilient yet strangely fragile natural environment to be protected at all costs.

Examples of pollutants reported to be in various beaches around the world, and the suspected source(s) of the pollutants.

LOCATION	POLLUTANT	SOURCE
Gaza Strip: Gaza Beach	Fecal coliform, fecal strepto-cocci, *Salmonella*, *Vibrio*, and/or *Pseudomonas*	Sewage from point and nonpoint sources
Israel: Bat Galim Beach, Haifa; Gordon Beach, Tel Aviv	Unspecified microbes, *Campylobacter jejuni*	Sewage
Portugal (33 beaches—four north; four central; nine Lisbon and Tagus Valley; one Alentejo; 15 Algarve)	Enterococci (ENT); *E. coli*; pathogenic fungi (most frequently *Candida & Aspergillus*); yeasts; dermatophytes; coliforms	Not specified
Brazil: Beaches of Olinda, Bairro Novo, Casa Caiada	Enteroparasites (nematodes), *Candida, Rhodotorula, Brettanomyces, Trichosporon*	Possibly from leaching or pluvial (rain) water and domestic wastes released along the beaches
France: Brittany beaches	Toxic green algae	Agriculture and fertilizers
Marseille	Bioavailable toxic contaminants (PAHs: polycyclic aromatic hydrocarbons; dioctyldiphenylamine)	Possibly car and truck emissions, industrial activities using fuel oils, coal and/or coke (iron and steel industries)

LOCATION	POLLUTANT	SOURCE
Corsica—La Marana	Bioavailable toxic contaminants (PAHs; dioctyldiphenylamine)	Possibly car and truck emissions, industrial activities using fuel oils, coal and/or coke (iron and steel industries)
New Zealand: Auckland, Narrow Neck Beach	Toxic algal blooms	Excess nutrients from farms
Italy: Genoa, La Spezia	*Ostreopsis ovata* (toxic algae)	Global warming; nitrate-packed fertilizers
Guadeloupe and *St. Kitts*	Larvae of hookworms; eggs of *Toxocara* spp.	Animal feces
England (two northwest coast beaches; two southwest coast beaches)	Pathogenic *Campylobacter jejuni, Salmonella*	Not specified
West Coast (U.S.) (55 beaches between Mexico and Oregon)	ENT, *E. coli*	Human waste
California (U.S.):		
Lovers Point, Monterey	Fecal indicator bacteria	Human waste
Mother's Beach, Marina del Rey	ENT	Sheltered (pocket) beach area
Cabrillo Beach, San Pedro	ENT	Sheltered (pocket) beach area
Santa Monica State Beach, near pier	ENT (levels approx. 1,000 times higher than those observed at open-ocean beaches)	Sheltered (pocket) beach area
Topanga Beach, Malibu	ENT (levels approx. 1,000 times higher than those observed at open-ocean beaches)	Sheltered (pocket) beach area
Avalon, Doheny, Malibu Surfrider	*Staphylococcus aureus*, MRSA, ENT	Not specified
Florida (U.S.):		
Hollywood Beach	Fecal bacteria (*E. coli*, ENT)	Sewage, animal waste
Ft. Lauderdale, Hollywood Beach, Hobie Beach	Fecal coliforms, *E. coli*, ENT, somatic coliphages, F+ coliphages, various yeasts (most frequently *Candida tropicalis, Rhodotorula mucilaginosa*)	Gull feces, other

LOCATION	POLLUTANT	SOURCE
Hobie Cat Beach, Key Biscayne, Miami	*Candida tropicalis, Cryptosporidium* spp., *Enterovirus, Staphylococcus aureus,* nematode larvae	Nonpoint source
Illinois (U.S.): 63rd St. Beach, Chicago, Lake Michigan	*E. coli, Salmonella enterica* serovar Typhimurium	Gulls, bacteria re-population
Michigan (U.S.): Grand Traverse Bay	ENT	Not specified
Southern Lake Michigan beaches	*Cladophora, E. coli*, ENT	Nonpoint sources or nearby inputs
Washington state (U.S.) (10 public beaches)	*Staphylococcus* spp., MRSA	Possibly swimmers
Wisconsin (U.S.): North Beach, Racine	*E. coli*, ENT, *Salmonella, Campylobacter*	Gull feces
Massachusetts (U.S.): Lake Attitash	Cyanobacteria	Not specified
Hawaii (U.S.): Hanauma Bay	Fecal coliform, *E. coli*, ENT	Pigeons implicated (highest concentrations of bacteria in dry sands where people congregate to sunbathe and eat)
Rhode Island (U.S.): Goddard Memorial State Park Beach	Fecal bacteria, Bacteroidales, ENT, *Clostridium*, F+ coliphage	Waste-treatment outfall
Alabama (U.S.): Fairhope Municipal Park Beach	Fecal bacteria, Bacteroidales, ENT, *Clostridium*, F+ coliphage	Waste-treatment outfall

Examples of beach pollutants and associated illnesses. This is a skeletal list of only a small number of beach pollutants.

AGENT	POSSIBLE MEDICAL OUTCOMES
INDICATOR BACTERIA (indicating fecal pollution and possible presence of enteric pathogens)	
Escherichia coli (*E. coli*; fecal coliform)	Gastroenteritis
Enterococcus (ENT)	Urinary-tract, wound, and soft-tissue infections; bacteremia
Fecal *Streptococci*	Strep infections
Bacteroidales	Gastrointestinal illnesses
INDICATOR VIRUSES (indicating fecal pollution and the possible presence of enteric pathogens)	
Coliphages (viruses that reproduce in *E. coli* bacteria)	Gastroenteritis
Somatic coliphages, F+ coliphages	Gastroenteritis, hepatitis, poliomyelitis, respiratory illness
PATHOGENIC BACTERIA (possibly causing illness or disease)	
Pseudomonas aeruginosa (widespread in the environment)	Acute respiratory distress syndrome, pneumococcal infections, various pneumonias, bacterial sepsis, urinary-tract infections, GI and soft-tissue infections, otitis externa

AGENT	POSSIBLE MEDICAL OUTCOMES
Salmonella and *Salmonella enterica*, serovar Typhimurium (in animal feces)	Gastroenteritis, typhoid fever, food poisoning, septicemia (generalized infections)
Staphylococcus aureus (in skin and nose, beach sand and water); MRSA; MRCoNS	Staph infections of skin, toxic shock syndrome, impetigo, food poisoning, endocarditis, heart failure, osteomyelitis, cellulitis, scalded-skin syndrome, bacteremia or sepsis, pneumonia, inflammation of the bones (osteomyelitis), circulatory collapse, death
Campylobacter jejuni (in feces)	Gastroenteritis
Clostridium (in sewage, feces)	Food poisoning, necrotizing enteritis (pigbel); rare in the United States; may cause death from infection and necrosis of intestines and the resulting septicemia
Vibrio spp. (in estuarine and marine waters)	Cholera, skin and tissue infection, death (in those with liver problems), gastrointestinal illnesses (diarrhea), sepsis (rare)

FUNGI AND YEASTS

Candida tropicalis, *Candida albicans*, and *Candida* spp. (normal in human body, in feces, and widespread in the environment)	Yeast infections, bloodstream infections (candidaemia), tissue-invasive candidiasis (less common)
Torulopsis (normal in human body)	Septicemia, vaginal infections, meningitis, pneumonia, bladder pain, pulmonary infections, urinary-tract infections
Rhodotorula mucilaginosa (common in the environment)	Fungemia, sepsis, endophthalmitis, peritonitis, meningitis (possible in immunocompromised people)
Trichosporon spp. (widespread, including in hair, soil, seawater, parrot droppings)	Trichosporonosis (in immunocompromised people), endocarditis, fungemia, infections
Black yeasts (*Exophiala dermatitidis*, *Aureobasidium pullulans*; can occur in healthy people)	Injury-associated localized infections (rare, but invasive infections) (in immunocompromised people), mycoses (lungs, skin), central nervous-system infections

PARASITES

Nematodes (roundworms; *Toxocara* spp.; in animal feces on the beach)	Parasitic nematodes (found in the intestines, muscles, and other tissues; more people around the world have a nematode infection than any other parasitic infection—either from ingesting eggs or through the skin via a wound)
Ankylostoma larvae (hookworms, cutaneous larva migrans; in animal feces on beach; the pathway to infection is walking barefoot on the beach or sitting on sand without a towel)	Anemia, lung symptoms (wheezing, coughing), diarrhea, bloody stool, abdominal pain, weight loss, loss of appetite

AGENT	POSSIBLE MEDICAL OUTCOMES
Cryptosporidium spp. (in human and animal feces)	Gastroenteritis, watery diarrhea, fever, abdominal cramps, nausea, vomiting

OTHER ORGANISMS

Ostreopsis ovata (marine dinoflagellate, a toxic algae)	Respiratory problems, conjunctivitis, dermatitis, body-temperature alterations, food poisoning (after consuming contaminated fish)
Cyanobacteria (blue-green algae, not related to other algae)	Can affect liver or central nervous system or cause gastrointestinal and kidney problems
Red tide (marine algae) and saxitoxins	Neurotoxic shellfish poisoning; eye irritation, sore throat

TOXIC CHEMICAL POLLUTANTS
(Nonspecific chemicals from melting plastic)

PAHs (polycyclic aromatic hydrocarbons)	Pulmonary, renal, gastrointestinal, dermatologic problems; various cancers (e.g., lung, skin, and bladder carcinoma)
Dioctyldiphenylamine (in hydrocarbons)	Eye irritation, skin disorders (e.g., dermatitis, folliculitis), irritation of nose, throat, and upper respiratory tract

Chapter 1. The End Is Nigh!

"AlHendon." "Great Resort but Disappearing Beaches." Review of Iberostar Vara-
dero, Cuba. TripAdvisor.com, June 28, 2008. http://www.tripadvisor.com
/ShowUserReviews-g147275-d535717-r17302079-Iberostar_Varadero-Varadero
_Matanzas_Province_Cuba.html.

Aquino, A. "Lanikai Beach Is Disappearing." YouTube, June 14, 2011. http://www
.youtube.com/watch?v=6_4mRGya2XA.

Aurofilio. "The Disappearing Beach Dilemma." *Auroville Today*, November 2002.
http://www.auroville.org/journals&media/avtoday/archive/2000-2003/Nov
_2002/beach.htm.

Barcelona Field Studies Centre. "Coastal Management: Sitges Case Study." 2013.
http://geographyfieldwork.com/CoastalManagementSitges.htm.

Bosker, G., and L. Lenceck. *Beaches*. San Francisco: Chronicle Books, 2000.

Coastalcare.org. http://coastalcare.org/.

Cooper, J. A. G., J. McKenna, D. W. T. Jackson, and M. O'Connor. "Mesoscale
Coastal Behavior Related to Morphological Self-adjustment." *Geology* 35
(2007): 187–90.

Defeo, O., A. McLachlan, D. S. Schoeman, T. A. Schlacher, J. Dugan, A. Jones,
M. Lastra, and F. Scapini. "Threats to Sandy Beach Ecosystems: A Review."
Estuarine, Coastal and Shelf Science 81 (2009): 1–12.

Delestrac, D., dir. *Sand Wars*. Paris: Rappi Productions and La compagnie des
Taxi-Brousse, 2013.

European Environment Agency. "The Squeeze on Europe's Coastline Continues."

Press release, November 28, 2013. http://www.eea.europa.eu/media/news releases/the-squeeze-on-europe2019s-coastline-continues/.

Gayes, P. T. "Post-Hurricane Hugo Side-Scan Sonar Survey: Impacts to Nearshore Morphology." In O. H. Pilkey and C. Finkle, eds., *The Impacts of Hurricane Hugo-September 10–22, 1989. Journal of Coastal Research Special Issue 8* (1991): 95–113.

Gornitz, V. *Rising Seas: Past, Present, Future*. New York: Columbia University Press, 2013.

Green, A. N., J. A. G. Cooper, R. Leuci, and Z. Thackeray. "Formation and Preservation of an Overstepped Segmented Lagoon Complex on the South African Continental Shelf. *Sedimentology* 60 (2013): 1755–68.

Grimsby Telegraph. "Our Beach Needs Better Protection." September 16, 2011. http://www.grimsbytelegraph.co.uk/beach-needs-better-protection/story -13343327-detail/story.html.

Harkinson, J. "EPA Scientist Says East Coast Beaches Threatened by Sea Level, but Nobody's Listening." *Wired*, in collaboration with *Mother Jones*, April 27, 2010. http://www.wired.com/2010/04/climate-desk-sea-level/.

Hayden, B. P., and N. R. Hayden. "Decadal and Century-Long Changes in Storminess at Long-Term Ecological Research Sites." In *Climate Variability and Ecosystem Response at Long-Term Ecological Research Sites*, ed. D. Greenland, D. G. Goodin, R. C. Smith, 262–85. New York: Oxford University Press, 2003.

Hine, A. C., R. B. Halley, S. D. Locker, B. D. Jarrett, W. C. Jaap, D. J. Mallinson, K. T. Ciembronowicz, N. B. Ogden, B. T. Donahue, and D. F. Naar. "Coral Reefs, Present and Past, on the West Florida Shelf and Platform Margin." In *Coral Reefs of the USA*, ed. B. M. Riegl and R. E. Dodge, 127–73. New York: Springer, 2008.

Hobbs, C. *The Beach Book: Science of the Shore*. New York: Columbia University Press, 2012.

Hughes, H., and J. Duchaine. *Frommer's 500 Places to See before They Disappear*. New York: Wiley, 2011.

Japan Update. "Okinawa's Natural Beaches Disappearing at Rapid Rate." October 11, 2004. http://www.japanupdate.com/?id=997.

Jarrett, D., A. C. Hine, R. B. Halley, D. F. Naar, S. D. Locker, A. C. Neumann, D. Twichell, C. Hu, B. T. Donahue, W. C. Jaap, D. Palandro, and K. Ciembronowicz. "Strange Bedfellows—a Deep-Water Hermatypic Coral Reef Superimposed on a Drowned Barrier Island; Southern Pulley Ridge, SW Florida Platform Margin." *Marine Geology* 214 (2005): 295–307.

Kalina, B., producer and director. *Shored Up*. Philadelphia: Mangrove Media LLC, 2013.

Kaufman, W., and O. H. Pilkey. *The Beaches Are Moving: The Drowning of America's Shoreline*. Durham, NC: Duke University Press, 1983.

Kelley, J. T., D. F. Belknap, and S. Claesson. "Drowned Coastal Deposits with Associated Archaeological Remains from a Sea-Level 'Slowstand': Northwestern Gulf of Maine, USA." *Geology* 38, no. 8 (August 2010): 695–98.

Kelley, J. T., D. F. Belknap, A. R. Kelley, and S. H. Claesson. "A Model for Drowned Terrestrial Habitats with Associated Archeological Remains in the Northwestern Gulf of Maine, USA." *Marine Geology* 338 (2013): 1–16.

Kimball, S. *Where's the Beach? The Hows and Whys of Our Disappearing Beaches—Severe Storms, Changing of the Sea Level, and Shifting Sand.* Video, Jefferson Lab Science Series, October 25, 1990. http://education.jlab.org/scienceseries /beach.html.

Kington, T. "Italy's Elite Are Dismayed by Vanishing Beaches." *Guardian*, July 9, 2011. http://www.guardian.co.uk/world/2011/jul/10/italy-beaches-erosion -climate-change.

Kyle, R., W. D. Robertson, and S. L. Birnie. "Subsistence Shellfish Harvesting in the Maputaland Marine Reserve in Northern KwaZulu-Natal, South Africa: Sandy Beach Organisms." *Biological Conservation* 82 (1997): 173–82.

Lajovic, V. "Erosion Threatening to Destroy One of the Nicest Beaches in Budva." Visit-Montenegro.com, April 4, 2010. Accessed July 8, 2013. http://www.visit -montenegro.com/article-mne-22318.htm.

Lenceck, L., and G. Bosker. *The Beach: The History of Paradise on Earth*. New York: Viking, 1998.

Liew, S. C., A. Gupta, P. P. Wong, L. K. Kwoh. "Recovery from a Large Tsunami Mapped Over Time: The Aceh Coast, Sumatra." *Geomorphology* 114 (2010): 520–29.

Lighty, R. G., I. G. MacIntyre, and R. Stuckenrath. "Submerged Early Holocene Barrier Reef South-East Florida Shelf." *Nature* 275 (1978): 59–60.

Living Heritage. "Mapua School: Mapua—The Good, the Bad, and the Ugly; Disappearing Beaches of Mapua." Living Heritage, Tikanga Tuku Iho, 2013. http://www.livingheritage.org.nz/schools/primary/mapua/good-bad-ugly /beaches.php.

Loureiro C., Ó. Ferreira, and J. A. G. Cooper. "Extreme Erosion on High-energy Embayed Beaches: Influence of Megarips and Storm Grouping." *Geomorphology* 139–40 (2011): 155–71.

Makpol, M. "Government Mulls 600 Million Baht Plan to Prevent Pattaya Beach Disappearing in 5 Years." *Pattaya Mail*, January 27, 2011. http://www.pattaya mail.com/k2/government-mulls-600-million-baht-plan-to-prevent-pattaya -beach-disappearing-in-5-years-1200.

Mamu, M. "Our Disappearing Beaches." *Solomon Star News* (Honiara), March 29, 2010. http://solomonstarnews.com/viewpoint/private-view/4328--our -disappearing-beaches.

Mapes, L. W. "Elwha: The Grand Experiment to Tear Down Two Dams and Return an Olympic Wilderness to Its Former Glory." *Seattle Times*, September 17, 2011.

McLachlan, A. "Physical Factors in Benthic Ecology: Effects of Changing Sand Particle Size on Beach Fauna." *Marine Ecology Progress Series* 131 (1996): 205–17.

Mihaescu, D. "Officials: Romanian Black Sea Beaches Disappearing." Associated Press, May 17, 2010. http://www.boston.com/news/world/europe/articles/2010/05/17/officials_romanian_black_sea_beaches_disappearing/.

Morris, L. "Disappearing Beaches in Gambia." *Grist*, October 30, 2009. http://grist.org/climate-energy/disappearing-beaches-in-gambia/.

Morton, R. A. "Historical Shoreline Changes and Their Causes, Texas Gulf Coast." *Bureau of Economic Geology Geological Circular 77*, no. 6 (1977): 352–64.

Morton, R. A., and M. J. Pieper. "Shoreline Changes on Mustang Island and North Padre Island (Aransas Pass to Yarborough Pass)." *Bureau of Economic Geology Geological Circular 77*, no. 1 (1977).

Morton, R. A., and M. J. Pieper. "Shoreline Changes on Central Padre Island (Yarborough Pass to Mansfield Channel)." *Bureau of Economic Geology Geological Circular 77*, no. 2 (1977).

Munisamy, R. L. "Disappearing Beaches." We Love Mauritius (WeLuvMu). Accessed May 28, 2013. http://welovemauritius.org/node/5.

NDTV. "Save the Beach Campaign: Save India's Beaches." Video, NDTV Convergence, 2012. http://www.ndtv.com/convergence/ndtv/new/Ndtv-Show-Special.aspx?ID=169.

Noriega, R., T. A. Schlacher, and B. Smeuninx. "Reductions in Ghost Crab Populations Reflect Urbanization of Beaches and Dunes." *Journal of Coastal Research* 28, no. 1 (2012): 123–31.

Panorama (Gibraltar). "Gibraltar's Fast Disappearing Beaches." February 15, 2011. http://www.panorama.gi/localnews/headlines.php?action=view_article&article=7065&offset=0.

Pearce, V. "Sand Disappearing from Armier Beach." TimesofMalta.com, August 5, 2008. http://www.timesofmalta.com/articles/view/20080805/letters/sand-disappearing-from-armier-beach.219510.

Peterson, C. H., M. J. Bishop, G. A. Johnson, L. M. D'Anna, L. M. Manning. "Exploiting Beach Filling as an Unaffordable Experiment: Benthic Intertidal Impacts Propagating upwards to Shorebirds." *Journal of Experimental Marine Biology and Ecology* 338 (2006): 205–21.

Peterson, C. H., H. C. Summerson, E. Thomson, H. S. Lenihan, J. Grabowski, L. Manning, F. Michell, and G. Johnson. "Synthesis of Linkages between Benthic and Fish Communities as a Key to Protecting Essential Fish Habitat." *Bulletin of Marine Science* 66, no. 3 (2000): 759–74.

Pilkey, O. H., and K. C. Pilkey. *Global Climate Change: A Primer.* Durham, NC: Duke University Press, 2011.

Pilkey, O. H., and R. Young. *The Rising Sea.* Washington, DC: Island Press, 2009.

Price, F., and A. Spiess. "A New Submerged Prehistoric Site and Other Fishermen's Reports Near Mount Desert Island." *Maine Archaeological Society Bulletin* 47, no. 2 (2007): 21–35.

Schlacher, T. A., J. Dugan, D. S. Schoeman, M. Lastra, A. Jones, F. Scapini, A. McLachlan, and O. Defeo. "Sandy Beaches at the Brink." *Diversity and Distributions* 13 (2007): 556–60.

Schrader, R. "Mui Ne, Vietnam's Disappearing Beach." *Leave Your Daily Hell*, January 25, 2011. http://leaveyourdailyhell.com/2011/01/25/mui-ne-vietnams-disappearing-beach/.

Seabrook, J. "The Beach Builders: Can the Jersey Shore Be Saved?" *New Yorker*, July 22, 2013. http://archives.newyorker.com/?i=2013-07-22#folio=042.

Shareef, N. M. "Disappearing Beaches of Kerala." *Current Science* 92, no. 2 (January 25, 2007): 157–58. http://www.iisc.ernet.in/currsci/jan252007/157a.pdf.

Shepard, F. P. *Submarine Geology*. 2nd edn. New York: Harper & Row, 1963.

Short, A., and B. Farmer. *101 Best Australian Beaches*. Sydney: NewSouth Publishing, 2012.

Siripong, A. "The Beaches Are Disappearing in Thailand." Paper presented at the International Symposia on Geoscience Resources and Environments of Asian Terranes (GREAT 2008), 4th IGCP, 5th APSEG, Bangkok, Thailand, November 24–26, 2008.

Smith, A., A. Mather, L. Guastella, J. A. G. Cooper, P. J. Ramsay, and A. Theron. "Contrasting Styles of Swell-driven Coastal Erosion: Examples from KwaZulu-Natal, South Africa." *Geological Magazine* 147 (2010): 940–53.

Springer, A. "Climate Change Eroding France's Tourism Hubs: Government Study." *Die Welt*, June 5, 2009. http://www.welt.de/english-news/article3685705/Climate-change-eroding-Frances-tourism-hubs.html.

Stevenson, M. "UN Climate Change Conference in Mexico Highlights Cancun's Disappearing Beaches." *Huffington Post*, November 30, 2010. http://www.huffingtonpost.com/2010/11/30/un-climate-change-confere_n_790129.html.

Woodruff, B., et al. "The Problem of Disappearing Beaches." *Ecos*, no. 19 (February 1979): 26–31.

Wright, L. D., and A. D. Short. "Morphodynamic Variability of Surf Zones and Beaches: A Synthesis." *Marine Geology* 56 (1984): 93–118.

Chapter 2. Selling the Family Silver: Beach-Sand Mining

All Africa Global Media. "Liberia: Illicit Sand Mining at Night." *Inquirer* (Monrovia), January 3, 2013. http://allafrica.com/stories/201301030759.html.

Asia Beat. "'Lazy' Malaysian Sand 'Better Off in Singapore.'" January 10, 2012. Accessed February 7, 2014. http://asiabeat.wordpress.com/2012/01/10/lazy-malasian-sand-better-off-in-Singapore/.

Billings, M. "San Francisco Bay Sand Mining Raises Questions about Beach Erosion." *Examiner* (San Francisco), January 12, 2011. http://www.sfexaminer.com/sanfrancisco/san-francisco-bay-sand-mining-raises-questions-about-beach-erosion/Content?oid=2318082.

Brown, D. "Facing Tough Times, Barbuda Continues Sand Mining Despite Warnings." Inter Press Service News Agency, June 22, 2013. Accessed March 11, 2014.

http://www.ipsnews.net/2013/06/facing-tough-times-barbuda-continues
-sand-mining-despite-warnings/.

Butler, R. "'Massive Room for Fraud' in Barbuda Sand Mining." *Antigua Observer*,
June 20, 2013.

Byrnes, M. R., M. R. Hammer, T. D. Thibaut, and D. B. Snyder. "Effects of Sand
Mining on Physical Processes and Biological Communities Offshore New Jersey." *Journal of Coastal Research* 20, no. 1 (2004): 25–43.

Cambers, G. "A Viable Solution to Beach Sand Mining? Montserrat." Wise Coastal
Practices for Sustainable Human Development Forum, UNESCO, September
15, 1999. http://www.csiwisepractices.org/?read=88.

CDE Global. "Manufactured Sands—A Viable Alternative." 2013. http://www
.cdeglobal.com/newsletters/205/manufactured-sands-a-viable-alternative
?gclid=CMDGq_v3_LkCFZPItAodRnoALA.

Dean, C. "Next Victim of Warming: The Beaches." *New York Times*, June 20, 2006.
http://select.nytimes.com/gst/abstract.html?res=F10717FD35550C738EDDAF0
894DE404482&fta=y&incamp=archive:article_related.

Dredging News Online. "Singapore Accused of Launching 'Sand War.'" February
15, 2010. http://www.sandandgravel.com/news/article.asp?v1=12585.

Goa News. "No control on Illegal Sand Mining." *Navhind Times* (Panaji, Goa).
November 17, 2012. http://www.navhindtimes.in/goa-news/no-control-illegal
-sand-mining.

Gornitz, V. *Rising Seas: Past, Present, Future*. New York: Columbia University
Press, 2013.

Government Information Service. "Sand Mining Areas Identified." St. George's,
Grenada, July 12, 2013. http://www.gov.gd/egov/news/2013/jul13/12_07_13
/item_1/sand_mining_areas_identified.html.

Gray, D. D. *Sand Mines Boom in Asia—at a Cost to Nature*. NBC News, August 22,
2011. http://www.nbcnews.com/id/44230562/#.UVCpgRkxt_k.

Gray, D. D. "Sand Mining Puts Nations' Environments at Risk." *San Francisco
Chronicle*, September 4, 2011.

Henderson, B. "Singapore Accused of Launching 'Sand Wars.'" *Telegraph*, February 12, 2010. http://www.telegraph.co.uk/news/worldnews/asia/singapore
/7221987/Singapore-accused-of-launching-Sand-Wars.html.

Hilton, M. J. "Applying the Principle of Sustainability to Coastal Sand Mining:
The Case of Pakiri-Mangawhai Beach, New Zealand." *Environmental Management* 18, no. 6 (1994): 815–29.

IRIN. "Sand-Mining Threatens Homes and Livelihoods in Sierra Leone." IRIN,
UN Office for the Coordination of Humanitarian Affairs, Nairobi, Kenya, February 4, 2013. http://reliefweb.int/report/sierra-leone/sand-mining-threatens
-homes-and-livelihoods-sierra-leone.

Isaac, C. "Sand Mining in Grenada: Issues, Challenges and Decisions Relating to
Coastal Management." Presentation at the Integrated Framework for the Management of Beach Resources within the Smaller Caribbean Islands workshop,

Mayaguez, Puerto Rico, October 21–25, 1996. UNESCO, CSI papers 1. http://
www.unesco.org/csi/pub/papers/abstra10.htm.

Jones, B. "Beach Mining Study Bodes Well for Prospectors: Washington State
Director Optimistic about Pilot Program." Gold Prospectors Association of
America, January 3, 2011. http://www.goldprospectors.org/Communication
/ArticlesandInformation/tabid/153/EntryId/246/Beach-mining-study-bodes
-well-for-prospectors.aspx.

Kanu, R. "How Illegal Sand Mining in Sierra Leone Is Destroying the Local
Beaches." *Ecologist*, April 3, 2013. http://www.theecologist.org/News/news
_analysis/1872134/how_illegal_sand_mining_in_sierra_leone_is_destroying
_the_local_beaches.html.

Lacey, M. "A Battle as the Tide Takes away Cancun Sand." *New York Times*, August
18, 2009. http://www.nytimes.com/2009/08/18/world/americas/18cancun.html.

Levitt, T. "The Damage Caused by Singapore's Insatiable Thirst for Land." *Ecologist*, May 11, 2010. http://www.theecologist.org/News/news_analysis/481729
/the_damage_caused_by_singapores_insatiable_thirst_for_land.html.

Mahima Groups. "Sand Mining Mafia Adopts New Tactics: Kozhikode, Kerala—
River Sand Mining and Export." *Times of India*, February 2, 2013. http://
timesofindia.indiatimes.com/city/kozhikode/Sand-mining-mafia-adopts
-new-tactics/articleshow/18265414.cms?referral=PM.

McLeod, M. "Pushing Grenada Backwards with Beach Sand Mining." *Grenada
Broadcast*, May 25, 2013. http://grenadaactionforum.com/2013/05/24/pushing
-grenada-backwards-with-beach-sand-mining/#more-789.

Minister of Works. "Sand Mining." *Grenada Broadcast*, June 6, 2012. http://www
.grenadabroadcast.com/news/other/13707-public-announcement.

New Today (Grenada). "Sand Mining Returning to Grenada." April 28, 2013.
http://thenewtoday.gd/local-news/2013/04/28/sand-mining-returning
-grenada/.

Pereira, K. "Sand Mining: The High Volume–Low Value Paradox." *Coastal Care*,
October 20, 2012. http://coastalcare.org/2012/10/sand-mining-the-high
-volume-low-value-paradox/.

Pereira, K. "Sand Mining—The Unexamined Threat to Water Security." *India
Water Portal*, December 17, 2012. http://www.indiawaterportal.org/sites/india
waterportal.org/files/article_for_india_water_portal_-_sand_mining_17th
_dec_2012_.pdf.

Pilkey, O. H., and R. Young. *The Rising Sea*. Washington, DC: Island Press, 2009.

Pilkey, O. H., R. S. Young, J. Kelley, and A. D. Griffith. "Mining of Coastal Sand:
A Critical Environmental and Economic Problem for Morocco." White paper,
Program for the Study of Developed Shorelines, Western Carolina University,
2009.

Seafriends. "Mining the Sea Sand." 1998. http://www.seafriends.org.nz/oceano
/seasand.htm.

Seltenrich, N. "SF Bay Sand Mining Alarms Conservationists." *Examiner* (San

Francisco), December 15, 2012. http://www.sfgate.com/science/article/SF-Bay
-sand-mining-alarms-conservationists-4121440.php.

Sengbeh, D. K. "Liberia: 'Illegal' Sand Mining, Sales." *Informer* (Monrovia), February 27, 2013. http://allafrica.com/stories/201302271107.html.

Thornton, E. "Beach Erosion Caused by CEMEX Sand Mining in Marina: Monterey County; Conservation Issues of the Ventana Chapter." Sierra Club, January 2012. http://ventana.sierraclub.org/conservation/marina/sandMiningErosion.shtml.

Tupufia, L. "Anger over Sandmining." *Samoa Observer*, December 27, 2012. http://www.samoaobserver.ws/home/headlines/2625-anger-over-sandmining.

United Nations. "Application for Inclusion of Sand Mining in the Agenda of the Convention of Biodiversity, a New and Emerging Issue Relating to the Conservation and Sustainable Use of Biodiversity." Awaaz Foundation and Bombay Natural History Society, 2013.

Young, R., and A. Griffith. "Documenting the Global Impacts of Beach Sand Mining." *Geophysical Research Abstracts* 11 (2009): 11593.

Chapter 3. Indefensible: Hard Structures on Soft Sand

Anfuso, G., and J. Á. M. del Pozo. "Assessment of Coastal Vulnerability Through the Use of GIS Tools in South Sicily (Italy)." *Environmental Management* 43, no. 3 (2009): 533.

Clayton, K. M. *Coastal Processes and Coastal Management, Technical Report*. Cheltenham, Countryside Commission, 1993.

Crammond, R. H. "A Consulting Engineer's Perspective." In *Papers and Proceedings of the Surgeon General's Conference on Agricultural Safety and Health: Public Law 101–517, April 30–May 3, 1991, Des Moines, Iowa*, ed. M. L. Myers, R. F. Herrick, S. A. Olenchock, et al. U.S. Department of Health and Human Services publication number 92–105: 365–71, 1992.

del Pozo, J. Á.M., and G. Anfuso. "Spatial Approach to Medium-term Coastal Evolution in South Sicily (Italy): Implications for Coastal Erosion Management." *Journal of Coastal Research* 24, no. 1 (2008): 33–42.

Dredging News Online. "British Beaches 'Could Be Gone within 100 Years.'" August 8, 2003. http://www.sandandgravel.com/news/article.asp?1=5337.

Gornitz, V. *Rising Seas: Past, Present, Future*. New York: Columbia University Press, 2013.

Hoover, H. *The Memoirs of Herbert Hoover*. 3 vols. New York: Macmillan, 1951.

Jackson, C. W., D. M. Bush, and W. J. Neal. "Documenting Beach Loss in Front of Seawalls in Puerto Rico: Pitfalls of Engineering a Small Island Nation Shore." In *Pitfalls of Shoreline Stabilization: Selected Case Studies*, ed. J. A. G. Cooper and O. H. Pilkey, 53–71. Dordrecht: Springer, 2012.

Kalina, B., producer and director. *Shored Up*. Philadelphia: Mangrove Media LLC, 2013.

Kelley, J. T., O. H. Pilkey, and J. A. G. Cooper, eds. *America's Most Vulnerable Coastal Communities*. Geological Society of America, special paper 460, 2009.

Kinver, M. "UK Floods: Learning Lessons from Past Storm Surges." BBC News, December 5, 2013. http://www.bbc.co.uk/news/science-environment-25247134.

Kraus, N. C., and O. H. Pilkey, eds. "The Effects of Seawalls on the Beach." *Journal of Coastal Research Special Issue No. 4* (1988).

Love, B., and M. Gabbett. "Recife: One of the World's Top 10 Shark Infested Beaches." Green Global Travel, August 16, 2012. http://greenglobaltravel.com/2012/08/16/shark-week-recife-brazil/.

Mapes, L. W. "Dams' Removal Promises Unique Chance to Start Over on a Grand Scale." *Seattle Times*, September 7, 2011.

Mapes, L. W. "Elwha: The Grand Experiment to Tear Down Two Dams and Return an Olympic Wilderness to Its Former Glory." *Seattle Times*, September 17, 2011.

Matthews, E. R. *Coast Erosion and Protection*. 2nd ed. London: C. Griffin and Company, 2007.

Pilkey, O. H. "Presque Isle Breakwaters: Successful Failures?" In *Pitfalls of Shoreline Stabilization: Selected Case Studies*, ed. J. A. G. Cooper and O. H. Pilkey, 131–39. Dordrecht: Springer, 2012.

Pilkey, O. H., and J. A. G. Cooper. "'Alternative' Shoreline Erosion Control Devices: A Review." In *Pitfalls of Shoreline Stabilization: Selected Case Studies*, ed. J. A. G. Cooper and O. H. Pilkey, 187–214. Dordrecht: Springer, 2012.

Pilkey, O. H., and K. L. Dixon. *The Corps and the Shore*. Washington, DC: Island Press, 1996.

Pilkey, O. H., W. J. Neal, J. T. Kelley, and J. A. G. Cooper. *The World's Beaches: A Global Guide to the Science of the Shoreline*. Berkeley: University of California Press, 2011.

Pilkey, O. H., and L. Pilkey-Jarvis. *Useless Arithmetic: Why Environmental Scientists Can't Predict the Future*. New York: Columbia University Press, 2007.

Pilkey, O. H., and R. Young. *The Rising Sea*. Washington, DC: Island Press, 2009.

Rennie, Sir John. "Presidential Address to the Annual General Meeting, Institution of Civil Engineers." Minutes of the proceedings, Institution of Civil Engineers, vol. 4 (1845), 23–25.

Shaler, N. S. *Sea and Land: Features of Coasts and Oceans with Special Reference to the Life of Man*. London: Smith, Elder and Co., 1895.

Ward, E. M. *English Coastal Evolution*. London: Methuen and Co., 1922.

Chapter 4. Patch-Up Jobs: Beach Replenishment

Alvarez, L. "Where Sand Is Gold, the Reserves Are Running Dry." *New York Times*, August 24, 2013.

American Shore and Beach Preservation Association. *Coastal Voice* 80, no. 1 (2013).

Coburn, A. S. "Beach Nourishment in the United States." In *Pitfalls of Shoreline*

Stabilization: Selected Case Studies, ed. J. A. G. Cooper and O. H. Pilkey, 105–19. Dordrecht: Springer, 2012.

Coburn, A. S. *Beach Nourishment Database*. Program for the Study of Developed Shorelines, Western Carolina University, 2013. http://www.psds-wcu.org /beach-nourishment.html.

Crain, D. A., A. B. Bolten, and K. Bjorndal. "Effect of Beach Nourishment on Sea Turtles: A Review and Research Initiatives." *Restoration Ecology* 3 (1995): 95–104.

Dean, R. C. "Beach Nourishment: Theory and Practice." *World Scientific* (2003): 420 pages.

Defeo, O., A. McLachlan, D. S. Schoeman, T. A. Schlacher, J. Dugan, A. Jones, M. Lastra, and F. Scapini. "Threats to Sandy Beach Ecosystems: A Review." *Estuarine, Coastal and Shelf Science* 81 (2009): 1–12.

Dixon, K. L., and O. H. Pilkey Jr. "Summary of Beach Replenishment Experience on the U.S. Gulf of Mexico Shoreline." *Journal of Coastal Research* 7 (1991): 249–56.

Gornitz, V. *Rising Seas: Past, Present, Future*. New York: Columbia University Press, 2013.

Haddad, T. C., and O. H. Pilkey. "Summary of the New England Beach Nourishment Experience (1935–1996)." *Journal of Coastal Research* 14, no. 4 (1998): 1395–404.

Hamm, L., M. Capobianco, H. H. Dette, A. Lechuga, R. Spanhoff, and M. J. F. Stive. "A Summary of European Experience with Shore Nourishment." *Coastal Engineering* 47 (2002): 237–64.

Kalina, B., producer and director. *Shored Up*. Philadelphia: Mangrove Media LLC, 2013.

Leonard, L. A., T. D. Clayton, and O. H. Pilkey Jr. "An Analysis of Replenished Beach Design Parameters on U.S. East Coast Barrier Islands." *Journal of Coastal Research* 6 (1990): 15–36.

Leonard, L. A., K. L. Dixon, and O. H. Pilkey Jr. "A Comparison of Beach Replenishment on the U.S. Atlantic, Pacific, and Gulf Coasts." "Artificial Beaches," *Journal of Coastal Research Special Issue No. 6* (1990): 127–40.

Mulder, J. P. M., and P. K. Tonnon. "'Sand Engine': Background and Design of a Mega-Nourishment Pilot in the Netherlands." *Proceedings of 32nd Conference on Coastal Engineering, Shanghai, China, 2010. Coastal Engineering Proceedings* 1, no. 32 (2011). http://journals.tdl.org/icce/index.php/icce/article/viewFile /1454/pdf_357. doi:10.9753/icce.v32.management.35.

Parr, T., D. Diener, and S. Lacy. "Effects of Beach Replenishment on the Nearshore Sand Fauna at Imperial Beach, California." Fort Belvoir, VA: U.S. Army Corps of Engineers, Coastal Engineering Research Center, National Technical Information Service, Operations Division (1978). http://dx.doi.org/10.5962 /bhl.title.47745.

Peterson, C. H., D. H. M. Hickerson, and C. G. Johnson. "Short-Term Consequences of Nourishment and Bulldozing on the Dominant Large Invertebrates of a Sandy Beach." *Journal of Coastal Research* 16 (2000): 368–78.

Pilkey, O. H. "Another View of Beachfill Performance." *Shore and Beach* 60, no. 2 (April 1992): 20–25.

Pilkey, O. H. "Beach Nourishment: Not the Answer." *Business and Economic Review* 53, no. 2 (2007): 7–8.

Pilkey, O. H. "A Time to Look Back at Beach Replenishment: Editorial." *Journal of Coastal Research* 6 (1990): iii–vii.

Pilkey, O. H. "What I Did on My Summer Vacation." Conclusion to the "Beach Nourishment: Is It Worth the Cost?" dialog, NOAA Coastal Services Center, 2006. http://www.csc.noaa.gov/archived/beachnourishment/html/human/dialog/series1d.htm.

Pilkey, O. H., and A. Coburn. "Beach Nourishment: Is It Worth the Cost?—Perspective." Dialog, NOAA Coastal Services Center, 2006. http://www.csc.noaa.gov/archived/beachnourishment/html/human/dialog/series1a.htm.

Pilkey, O. H., and A. Coburn. "Beach Nourishment: It's a Good Investment—Critique." Dialogue, NOAA Coastal Services Center, 2006. http://www.csc.noaa.gov/archived/beachnourishment/html/human/dialog/series1b.htm.

Pilkey, O. H., and A. Coburn. "What You Know Can Hurt You: Predicting the Behavior of Nourished Beaches." In *Prediction: Science, Decision Making, and the Future of Nature*, ed. D. Sarewitz, R. A. Pielke Jr., and R. Byerly Jr., 159–84. Washington, DC: Island Press, 2000.

Pilkey, O. H., and J. A. G. Cooper. "'Alternative' Shoreline Erosion Control Devices: A Review." In *Pitfalls of Shoreline Stabilization: Selected Case Studies*, ed. J. A. G. Cooper and O. H. Pilkey, 187–214. Dordrecht: Springer, 2012.

Pilkey, O. H., and K. L. Dixon. *The Corps and the Shore*. Washington, DC: Island Press, 1996.

Pilkey, O. H., and K. C. Pilkey. *Global Climate Change: A Primer*. Durham, NC: Duke University Press, 2011.

Pilkey, O. H., and L. Pilkey-Jarvis. *Useless Arithmetic: Why Environmental Scientists Can't Predict the Future*. New York: Columbia University Press, 2007.

Pilkey, O. H., and R. Young. *The Rising Sea*. Washington, DC: Island Press, 2009.

Rijkswaterstaat and Provincie Zuid-Holland. "The Sand Engine." The Netherlands, 2013. http://www.dezandmotor.nl/en-GB/.

Sistermans, P., and O. Nieuwenhuis. *EUROSION Case Study, Isle Of Sylt: Isles Scheslwig-Holstein* (Germany). DHV Group, 2002. http://www.eurosion.org/shoreline/17sylt.html; ec.europa.eu/ourcoast/index.cfm?menuID=7&articleID=195.

Speybroeck, J., D. Bonte, W. Courtens, T. Gheskiere, P. Grootaert, J-P. Maelfait, M. Mathys, S. Provoost, K. Sabbe, E. W. M. Stienen, V. Van Lancker, M. Vincx, and S. Degraer. "Beach Nourishment: An Ecologically Sound Coastal Defence Alternative? A Review." *Aquatic Conservation—Marine and Freshwater Ecosystems* 16 (2006): 419–35.

Trembanis, A. C., O. H. Pilkey, and H. R. Valverde. "Comparison of Beach Nourishment along the U.S. Atlantic, Great Lakes, Gulf of Mexico, and New England Shorelines." *Coastal Management* 27 (1999): 329–40.

Trembanis, A. C., H. R. Valverde, and O. H. Pilkey. "Comparison of Beach Nourishment along the U.S. Atlantic, Great Lakes, Gulf of Mexico and New England Shorelines." *Journal of Coastal Research Special Issue No. 26* (1998): 246–51.

U.S. Army Corps of Engineers. *Shoreline Protection and Beach Erosion Control Study: Phase I: Cost Comparison of Shoreline Protection Projects of the US Army Corps of Engineers.* IWR report 94 PS 1 ("The Purple Report"), 1994.

Valverde, H. R., A. C. Trembanis, and O. H. Pilkey. "Summary of Beach Nourishment Episodes on the U.S. East Coast Barrier Islands." *Journal of Coastal Research* 15, no. 4 (1999): 1100–1118.

Witherington, B., S. Hirama, and A. Mosier. "Barriers to Sea Turtle Nesting on Florida Beaches: Linear Extent and Changes following Storms." *Just Cerfing* 2, no. 6 (2011): 17–20.

Chapter 5. The Plastisphere: Trash on the Beach

Aguiar, E. "Beach Trash a Relentless Tide." *Honolulu Advertiser*, January 18, 2007. http://the.honoluluadvertiser.com/article/2007/Jan/18/In/FP701180353.html.

Amos, A. F. "Pollution of the Ocean by Plastic and Trash." *Water Encyclopedia: Science and Issues*, 2011. http://www.waterencyclopedia.com/Po-Re/Pollution-of-the-Ocean-by-Plastic-and-Trash.html.

Amos, A. F. *Solid Waste Pollution on Texas Beaches: A Post-MARPOL Annex V Study.* Vol. 1. OCS Study MMS 93–0013. New Orleans: U.S. Department of the Interior, Minerals Management Service, Gulf of Mexico OCS Region, 1993.

Associated Press. "One Day's Haul of Beach Trash: 6 Million Pounds." NBCNews.com, April 15, 2008. http://www.nbcnews.com/id/24141483/ns/world_news-world_environment/t/one-days-haul-beach-trash-million-pounds/#.Ua9X7uuXJ_k.

Avery-Gomm, S., P. D. O'Hara, L. Kleine, V. Bowes, L. K. Wilson, and K. L. Barry. "Northern Fulmars as Biological Monitors of Trends of Plastic Pollution in the Eastern North Pacific." *Marine Pollution Bulletin* 64, no. 9 (September 2012): 1776–81.

Browne, M. A., P. Crump, S. J. Niven, E. L. Teuten, A. Tonkin, T. Galloway, and R. C. Thompson. "Accumulation of Microplastic on Shorelines Worldwide: Sources and Sinks." *Environmental Science and Technology* 45, no. 21 (2011): 9175–79.

Depledge, M. H., F. Galgani, C. Panti, I. Caliani, S. Casini, M. C. Fossi. "Plastic Litter in the Sea." *Marine Environmental Research* 92 (2013): 279–81.

Ebbesmeyer, C., and E. Scigliano. *Flotsametrics and the Floating World: How One Man's Obsession with Runaway Sneakers and Rubber Ducks Revolutionized Ocean Science.* New York: Smithsonian Books/HarperCollins, 2009.

Environment News Service. "FEMA Gives $18.1 Million for Texas Beach Cleanup after Ike: Austin, Texas." May 4, 2009. http://www.ens-newswire.com/ens /may2009/2009-05-04-095.html.

Frazier, I. "Form and Fungus." *The New Yorker*, May 20, 2013, 50–62.

Hohn, D. "Sea of Trash." *New York Times*, June 22, 2008. http://www.nytimes.com /2008/06/22/magazine/22Plastics-t.html?pagewanted=all&_r=0.

Kiessling, I., and C. Hamilton. *Marine Debris at Cape Arnhem Northern Territory, Australia:* WWF *Report Northeast Arnhem Land Marine Debris Survey 2000.* Sydney: WWF Australia, 2001.

Leitch, K. *Entanglement of Marine Turtles in Netting: Northeast Arnhem Land, Northern Territory Australia; Report to* WWF *(Australia).* Sydney: WWF Australia, 1999.

Natural Resources Defense Council. *Testing the Waters.* Annual Report, 2013.

Nurhayati, D. "Kuta Trash Disrupts Surfing." *Jakarta Post*, January 26, 2013. http://www.thebalidaily.com/2013-01-26/kuta-trash-disrupts-surfing.html.

Ocean Conservancy. *Tracking Trash: 25 Years of Action for the Ocean, 2011 Report.* 2011. http://act.oceanconservancy.org/pdf/Marine_Debris_2011_Report_OC .pdf.

Oigman-Pszczol, S. S., and J. C. Creed. "Quantification and Classification of Marine Litter on Beaches along Armação dos Búzios, Rio de Janeiro, Brazil." *Journal of Coastal Research* 23, no. 2 (2007): 421–28.

Sahagun, L. "An Ecosystem of Our Own Making Could Pose a Threat." *Los Angeles Times*, December 26, 2013. http://www.latimes.com/science/la-sci -plastisphere-20131228,0,811701.story#axzz2rbbfHCmH.

Santos, I. R., A. C. Friedrich, and F. P. Barretto. "Overseas Garbage Pollution on Beaches of Northeast Brazil." *Marine Pollution Bulletin* 50 (2005): 778–86.

Winarti, A. "Flood of Trash Remains a Headache." *Jakarta Post*, March 5, 2012. http://www.thejakartapost.com/news/2012/03/05/flood-trash-remains-a -headache.html.

Zettler, E. R, T. J. Mincer, and L. A. Amaral-Zettler. "Life in the Plastisphere: Microbial Communities on Plastic Marine Debris." *Environmental Science and Technology* 47, no. 13 (2013): 7137–46.

Chapter 6. Tar Balls and Magic Pipes

Barker, J. M. "$3.5 Million Pollution Fine for Ship Operator." *Seattle Post- Intelligencer*, June 29, 2004. Accessed December 30, 2012. http://www.seattlepi .com/local/article/3-5-million-pollution-fine-for-ship-operator-1148302.php.

Bik, H. M., K. M. Halanych, J. Sharma, and W. K. Thomas. "Dramatic Shifts in Benthic Microbial Eukaryote Communities following the Deepwater Horizon Oil Spill." PLOS ONE 7, no. 6 (2012): e385550. http://www.plosone.org/article /info%3Adoi%2F10.1371%2Fjournal.pone.0038550.

Bishop, J. "Disaster in the Gulf: Go below the Surface of the Gulf Oil Disaster." Natural Resources Defense Council, n.d. Accessed December 12, 2012. http:// www.nrdc.org/energy/gulfspill/belowsurface.asp.

CBS / AP. "Isaac Churned up Old Oil from BP Spill in La., Tests Confirm."
CBSNews.com, September 6, 2012. Accessed December 10, 2012. http://www
.cbsnews.com/8301-201_162-57507123/isaac-churned-up-old-oil-from-bp
-spill-in-la-tests-confirm/.

del Nogal Sánchez, M., J. L. Pérez Pavón, M. E. Fernández Laespada, C. Garcia
Pinto, B. Moreno Cordero. "Factors Affecting Signal Intensity in Headspace
Mass Spectrometry for the Determination of Hydrocarbon Pollution in Beach
Sands." *Analytical and Bioanalytical Chemistry* 382 (2005): 372–80.

Department of Ecology, State of Washington. *Pre-booming Requirements for
Delivering Vessels*, 2007. http://www.ecy.wa.gov/programs/spills/prevention
/VesselTechAssist/Dvessel_prebooming.html.

Fiolek, A., L. Pikula, and B. Voss. *Resources on Oil Spills, Response and Restoration:
A Selected Bibliography*. U.S. Department of Commerce, National Oceanic and
Atmospheric Administration, revised 2011.

Georgia Institute of Technology. "Gulf of Mexico Clean-up Makes 2010 Spill
52-Times More Toxic: Mixing Oil with Dispersant Increased Toxicity to Ecosystems." *ScienceDaily*, November 30, 2012. Accessed December 30, 2012.
http://www.sciencedaily.com/releases/2012/11/121130110518.htm.

Goa News. "Beach Tar Balls Due to Tanker Oil Spills: Study." June 20, 2013. http://
www.goacom.com/goa-news-highlights/10282-beach-tar-balls-due-to-tanker
-oil-spills-study.

Hayes, M. *Black Tides*. Austin: University of Texas Press, 1999.

Hayes, M., et al. "The Gulf War Oil Spill Twelve Years Later: Long-Term Impacts
to Coastal and Marine Resources." Paper presented at the Offshore Arabia
Conference in Dubai. Mid-December Proceedings, 2006.

International Maritime Organization, 1973/78. *International Convention for the
Prevention of Pollution from Ships* (MARPOL). Adoption: 1973 (Convention),
1978 (1978 Protocol), 1997 (Protocol–Annex VI); Entry into force: 2 October
1983 (Annexes I and II). http://www.imo.org/KnowledgeCentre/References
AndArchives/HistoryofMARPOL/Pages/default.aspx.

Jones, D. A., M. Hayes, F. Krupp, G. Sabatini, I. Watt, and L. Weishar. "The Im-
pact of the Gulf War (1990–1991) Oil Release Upon the Intertidal Gulf Coast
Line of Saudi Arabia and Subsequent Recovery." In *Protecting the Gulf's
Marine Ecosystems from Pollution*, ed. A. A. Abuzinada, H-J. Barth, F. Krupp,
B. Böer, T. Z. Al Abdessalaam (2008), 237–54.

Kearsley, K. "Greek Shipping Co. Fined for Illegal Dumping." *News Tribune*
(Tacoma), June 28, 2007.

Kiern, L. "Manager of M/V Chem Faros to Plead Guilty to MARPOL Violations."
Winston & Strawn, May 12, 2010.

Leftwich, R. "Greek Shipping Company Fined Millions for Dumping Oil off US
Shores." ABC2News.com, September 21, 2010. Accessed July 10, 2013. http://
www.abc2news.com/dpp/news/national/greek-shipping-company-fined
-millions-for-dumping-oil-off-us-shores.

Michel, J., M.O. Hayes, C.D. Getter, and L. Cotsapas. "The Gulf War Oil Spill Twelve Years Later: Consequences of Eco-terrorism." Proceedings of the 2005 International Oil Spill Conference, American Petroleum Institute, 2005.

Neel, J., C. Hart, D. Lynch, S. Chan, and J. Harris. *Oil Spills in Washington State: A Historical Analysis*. Publication no. 97-252, Washington State Department of Ecology, Spill Management Program, revised 2007. https://fortress.wa.gov/ecy/publications/publications/97252.pdf.

NOAA's National Ocean Service. "Tarballs" and "Guide to Expected Forms of Oil: Deepwater Horizon Oil Spill." Fact sheet, Office of Response and Restoration, Emergency Response Division, n.d. http://www.dep.state.fl.us/deepwater horizon/files/tar_ball_info.pdf.

Paris, C. B., M. Le Hénaff, Z. M. Aman, A. Subramaniam, J. Helgers, D. P. Wang, V. H. Kourafalou, and A. Srinivasan. "Evolution of the Macondo Well Blow-out: Simulating the Effects of the Circulation and Synthetic Dispersants on the Subsea Oil Transport." *Environmental Science and Technology* 46 (2012): 13293–302.

Pérez-Cadahía, B., B. Laffon, V. Valdiglesias, E. Pásaro, and J. Méndez. "Cyto-genetic Effects Induced by Prestige Oil on Human Populations: The Role of Polymorphisms in Genes Involved in Metabolism and DNA Repair." *Mutation Research* 653, no. 1–2 (2008): 117–23.

Petrow, R. *In the Wake of Torrey Canyon*. New York: David McKay, 1968.

Plataforma SINC. "Hydrocarbon Pollution along the Coast of Galicia Shot Up Five Years after the Prestige Oil Spill." *ScienceDaily*, November 22, 2011. Accessed December 30, 2012. http://www.sciencedaily.com/releases/2011/11/111122112028.htm.

Politicol News. "World's Oil Spill List." May 30, 2010. Accessed December 30, 2012. http://www.politicolnews.com/worlds-oil-spill-list/.

Robertson, C., and C. Krauss. "Gulf Spill Is the Largest of Its Kind, Scientists Say." *New York Times*, August 2, 2010.

Schleifstein, M. "Greek Shipping Company Fined $1.2 million; Sentenced to 3 Years Probation for Dumping Oil Wastes from Ship Traveling to New Orleans." *Times-Picayune* (New Orleans), July 25, 2012. Accessed July 10, 2013. http://www.nola.com/environment/index.ssf/2012/07/greek_shipping _company_fined_1.html.

Symons, L. C. "NOAA's Remediation of Underwater Legacy Environmental Threats (RULET) Database & Wreck Oil Removal Program (WORP)." NOAA Office of National Marine Sanctuaries, slide show, forty-one slides, 2012. http://www.nrt.org/production/nrt/RRTHomeResources.nsf/resources /RRT4Feb2013Meeting_1/$File/RULET_RRTIV.pdf.

Tawfiq, N. F., and D. A. Olsen. "Saudi Arabia's Response to the 1991 Gulf Oil Spill." *Marine Pollution Bulletin* 27 (1993): 333–45.

U.S. Attorney's Office, Western District of Washington. "Marine Conservation Fund Awards $1.7 Million in Grants to Restore Puget Sound: Funding Comes from Settlement with Shipping Line That Used "Magic Pipe" to Illegally Dump Oil." July 5, 2006. Accessed December 30, 2012. http://www.justice.gov/usao /waw/press/2006/jul/marine.htm.

U.S. Department of Justice. "Ship Serial Polluter Ordered to Pay $4 Million for Covering up the Deliberate Discharge of Oil and Plastics." Press release, Office of Public Affairs, September 21, 2010. Accessed December 31, 2012. http://www .justice.gov/opa/pr/2010/September/10-enrd-1059.html.

Valdiglesias, V., G. Kilic, C. Costa, O. Amor-Carro, L. Mariñas-Pardo, D. Ramos-Barbón, J. Méndez, E. Pásaro, and B. Laffon. 2012. "In Vivo Genotoxity Assessment in Rats Exposed to Prestige-Like Oil by Inhalation." *Journal of Toxicology and Environmental Health* 75, nos. 13–15 (2012): 756–64.

Wang, P., and T. M. Roberts. "Distribution of Surficial and Buried Oil Contaminants across Sandy Beaches along NW Florida and Alabama Coasts Following the Deepwater Horizon Oil Spill in 2010." *Journal of Coastal Research* 29, no. 6a (2013): 144–55.

Chapter 7. Stuck in a Rut: Driving on the Beach

Anders, F. J., and S. P. Leatherman. "Disturbance of Beach Sediment by Off-Road Vehicles." *Environmental Geology and Water Science* 9 (1987): 183–89.

Anders, F. J., and S. P. Leatherman. "Effects of Off-Road Vehicles on Coastal Foredunes at Fire Island, New York, USA." *Environmental Management* 11 (1987): 45–52.

Basu, J. "On the Waterfront: This Seaside Resort Is a Short Hop from the City and a Shorter Hop away from Disaster." *Telegraph* (Calcutta), December 6, 2009. http://www.telegraphindia.com/1091206/jsp/calcutta/story_11828847.jsp.

Buick, A. M., and D. C. Paton. "Impact of Off-Road Vehicles on the Nesting Success of Hooded Plovers *Charadrius rubricollis* in the Coorong Region of South Australia." *Emu* 89 (1989): 159–72.

Celliers, L., T. Moffett, N. C. James, and B. Q. Mann. "A Strategic Assessment of Recreational Use Areas for Off-Road Vehicles in the Coastal Zone of KwaZulu-Natal, South Africa." *Ocean and Coastal Management* 47 (2004): 123–40.

Cox, J. H., H. F. Percival, and S. V. Colwell. *Impact of Vehicular Traffic on Beach Habitat and Wildlife at Cape San Blas, Florida*. Technical report 50. Gainesville, Florida Cooperative Fish and Wildlife Research Unit, 1994.

Davenport, J., and J. L. Davenport. "The Impact of Tourism and Personal Leisure Transport on Coastal Environments: A Review." *Estuarine, Coastal and Shelf Science* 67 (2006): 280–92.

Godfrey, P. J., and M. Godfrey. "Ecological Effects of Off-Road Vehicles on Cape Cod." *Oceanus* 23 (1981): 56–67.

Godfrey, P. J., S. Leatherman, and P. Buckley. "Impact of Off-Road Vehicles on Coastal Ecosystems." In *Proceedings of the Symposium on Technical, Environmental, Socio-economic and Regulatory Aspects of Coastal Zone Planning and Management*, San Francisco, 1978, 581–600.

Godfrey, P. J., S. Leatherman, and P. Buckley. "ORVs and Barrier Beach Degradation." *Parks*, no. 2 (1980): 5–11.

Goldin, M. R. "Effects of Human Disturbance and Off-Road Vehicles on Piping
Plover Reproductive Success and Behavior at Breezy Point, Gateway National
Recreation Area, New York." MS thesis, University of Massachusetts, Amherst,
1993.

Hosier, P. E., and T. E. Eaton. "The Impact of Vehicles on Dune and Grassland
Vegetation on a Southeastern North Carolina Barrier Beach." *Journal of Applied Ecology* 17 (1980): 173–82.

Hosier, P. E., M. Kochhar, and V. Thayer. "Off-Road Vehicle and Pedestrian Track
Effects on the Sea-Approach of Hatchling Loggerhead Turtles." *Environmental
Conservation* 8 (1981): 158–61.

Kluft, J. M., and H. S. Ginsberg. *The Effect of Off-Road Vehicles on Barrier Beach
Invertebrates at Cape Cod and Fire Island National Seashores.* Technical report,
NPS / NER / NRTR—2009 / 138, National Park Service, Boston, April 2009.

Kudo, H., A. Murakami, and S. Watanbe. "Effects of Sand Hardness and Human
Beach Use on Emergence Success of Loggerhead Sea Turtles on Yakushima
Island, Japan." *Chelonian Conservation and Biology* 4 (2003): 695–96.

Lamont, M., H. F. Percival, and S. V. Colewell. "Influence of Vehicle Tracks on
Loggerhead Hatchling Seaward Movement along a Northwest Florida Beach."
Florida Field Naturalist 30 (2002): 77–109.

Leatherman, S. P., and P. J. Godfrey. *The Impact of Off-Road Vehicles on Coastal
Ecosystems in Cape Cod National Seashore: An Overview.* Environmental Insti-
tute, University of Massachusetts, Amherst, National Park Service Cooperative
Research Unit, report no. 34, 1979.

Lutcavage, M. E., P. Plotkin, B. Witherington, and P. L. Lutz. "Human Impacts on
Sea Turtle Survival." In *The Biology of Sea Turtles*, vol. 1, ed. P. L. Lutz and J. A.
Musick, 387–411. Boca Raton, FL: CRC Press, 1997.

Mdletshe, C. "Beach Driving Fines Hiked." *Times LIVE*, March 14, 2013. http://
www.timeslive.co.za/thetimes/2013/03/14/beach-driving-fines-hiked.

Melvin, S. M., A. Hecht, and C. R. Griffin. "Piping Plover Mortalities Caused
by Off-Road Vehicles on Atlantic Coast Beaches." *Wildlife Society Bulletin* 22
(1994): 409–14.

Moss, D., and D. P. McPhee. "The Impacts of Recreational Four-Wheel Driving
on the Abundance of the Ghost Crab (*Ocypode cordimanus*) on Subtropical
Beaches in SE Queensland." *Coastal Management* 34 (2006): 133–40.

Nelson, D. A., and B. Blihovde, 1998. "Nesting Sea Turtle Response to Beach
Scarps." In *Proceedings of the Sixteenth Annual Symposium on Sea Turtle Bi-
ology and Conservation*, ed. R. Byles and Y. Fernandez, 113. NOAA Technical
Memorandum NMFS-SEFSC-412.

Nester, L. R. "Effects of Off-Road Vehicles on the Nesting Activity of Loggerhead
Sea Turtles in North Carolina." MSc. thesis, University of Florida, 2006.

Oregon Beach Bill. HB 1601, Chapter 601, An Act. *Oregon Laws and Resolutions:*

Enacted and Adopted by the Regular Session of the Fifty-fourth Legislative Assembly Beginning January 9 and Ending June 14, 1967. Salem, Oregon: Oregon Legislative Assembly, 1967. http://www.govoregon.org/beachbilltext.html.

Rickard, C. A., A. McLachlan, and G. I. H. Kerley. "The Effects of Vehicular and Pedestrian Traffic on Dune Vegetation in South Africa." *Ocean and Coastal Management* 23 (1994): 225–47.

Schlacher, T. A., J. Dugan, D. S. Schoeman, M. Lastra, A. Jones, F. Scapini, A. McLachlan, and O. Defeo. "Sandy Beaches at the Brink." *Diversity and Distributions* 13 (2007): 556–60.

Schlacher, T. A., D. Richardson, and I. McLean. "Impacts of Off-Road Vehicles (ORVs) on Macrobenthic Assemblages on Sandy Beaches." *Environmental Management* 41 (2008): 878–92.

Schlacher, T. A., and L. M. C. Thompson. "Exposure of Fauna to Off-Road Vehicle (ORV) Traffic on Sandy Beaches." *Coastal Management* 35 (2007): 567–83.

Schlacher, T. A., and L. M. C. Thompson. "Physical Impacts Caused by Off-Road Vehicles to Sandy Beaches: Spatial Quantification of Car Tracks on an Australian Barrier Island." *Journal of Coastal Research* 24 (2008): 234–42.

Schlacher, T. A., L. M. C. Thompson, and S. Price. "Vehicles versus Conservation of Invertebrates on Sandy Beaches: Quantifying Direct Mortalities Inflicted by Offroad Vehicles (ORVs) on Ghost Crabs." *Marine Ecology—Evolutionary Perspective* 28 (2007): 354–67.

Schlacher, T., L. M. C. Thompson, and S. Walker. "Mortalities Caused by Off-Road Vehicles (ORVs) to a Key Member of the Sandy Beach Assemblages, the Surf Clam *Donax Deltoides.*" *Hydrobiologia* 610 (2008): 345–50.

Shaw, D. "The Beach Is the Road, and the Commute Is an Adventure." *New York Times*, March 4, 2007. http://www.nytimes.com/2007/03/04/realestate/04Habi .html.

Sheppard, N., K. A. Pitt, and T. A. Schlacher. "Sub-lethal Effects of Off-Road Vehicles (ORVs) on Surf Clams on Sandy Beaches." *Journal of Experimental Marine Biology and Ecology* 380 (2009): 113–18.

Steiner, A. J., and S. P. Leatherman. *An Annotated Bibliography of the Effects of Off-Road Vehicle and Pedestrian Traffic on Coastal Ecosystems.* University of Massachusetts and National Park Service Cooperative Research Unit, report no. 45, 1979.

Steiner, A. J., and S. P. Leatherman. "Recreational Impacts on the Distribution of Ghost Crabs *Ocypode quadrata fab.*" *Biological Conservation* 20, no. 2 (1981): 111–22.

Stephenson, G. *Vehicle Impacts on the Biota of Sandy Beaches and Coastal Dunes: A Review from a New Zealand Perspective.* New Zealand Department of Conservation, Wellington, 1999. http://doc.org.nz/Documents/science-and -technical/sfc121.pdf.

Thompson, L. M. C., and T. A. Schlacher. "Physical Damage to Coastal Dunes and Ecological Impacts Caused by Vehicle Tracks Associated with Beach

Camping on Sandy Shores: A Case Study from Fraser Island, Australia." *Journal of Coastal Conservation* 12 (2008): 67–82.

U.S. Fish and Wildlife Service. "Endangered and Threatened Wildlife and Plants: Revised Designation of Critical Habitat for the Wintering Population of Piping Plover (*Charadrius melodus*) in North Carolina; Final Rule." *Federal Register* 73, no. 204 (2008): 62815–41.

van der Merwe, D., and D. van der Merwe. "Effects of Off-Road Vehicles on the Macrofauna of a Sandy Beach." *South African Journal of Science* 87 (1991): 210–13.

Watson, J. J. "Dune Breeding Birds and Off-Road Vehicles." *The Naturalist* 36, no. 3 (1992): 8–12.

Wheeler, N. R. *Effects of Off-Road Vehicles on the Infauna of Hatches Harbor, Cape Cod National Seashore, Massachusetts.* University of Massachusetts and National Park Service Cooperative Research Unit, report no. 28, 1979.

Wolcott, T. G., and D. Wolcott. "Impact of Off-Road Vehicles on Macroinvertebrates of a Mid-Atlantic Beach." *Biological Conservation* 29 (1984): 217–40.

Zaremba R., P. J. Godfrey, and S. P. Leatherman. *The Ecological Effects of Off-Road Vehicles on the Beach/Backshore Zone in Cape Cod National Seashore, Massachusetts.* University of Massachusetts and National Park Service Cooperative Research Unit, report no. 28, 1979.

Chapter 8. The Enemy Within: Beach Pollution

Abdelzaher, A. M., M. E. Wright, C. Ortega, H. M. Solo-Gabriele, G. Miller, S. Elmir, X. Newman, P. Shih, J. A. Bonilla, T. D. Bonilla, et al. "Presence of Pathogens and Indicator Microbes at a Non-point Source Subtropical Recreational Marine Beach." *Applied Environmental Microbiology* 76, no. 3 (2010): 724–32.

American Chemical Society. "Beach Sand May Harbor Disease-Causing E. Coli Bacteria." *ScienceDaily*, May 29, 2007. Accessed June 27, 2013, http://www.sciencedaily.com/releases/2007/05/070528095321.htm.

American Chemical Society. "New Insights into When Beach Sand May Become Unsafe for Digging and Other Contact." Phys.org, April 11, 2012. http://phys.org/news/2012-04-insights-beach-sand-unsafe-contact.html.

Ashour, F., B. Ashour, M. Komarzynski, Y. Nassar, M. Kudla, N. Shawa, and G. Henderson. *A Brief Outline of the Sewage Infrastructure and Public Health Risks in the Gaza Strip for the World Health Organisation.* Emergency Water and Sanitation/Hygiene, United Nations Information System on the Question of Palestine, April 2, 2009.

Associated Press. "After Spill, Waikiki Sand Is Clean, Health Group Says." *New York Times*, April 23, 2006. http://www.nytimes.com/2006/04/23/us/23Waikiki.html?_r=0.

BBC News. "France: Wild Boars Dead amid Algae on Brittany Coast." July 28, 2011. http://www.bbc.co.uk/news/world-europe-14324094.

Berger, T. "Creeping Nematodes." eHow.com, 2011. http://www.ehow.com /about_6457132_creeping-nematodes.html.

Bohan, S. "California 'Sand Pollution' Intrigues Researchers." redOrbit, September 12, 2007. http://www.redorbit.com/news/science/1063441/california_sand _pollution.

Bolton, F. J., S. B. Surman, K. Martin, D. R. Wareing, and T. J. Humphrey. "Presence of Campylobacter and Salmonella in Sand from Bathing Beaches." *Epidemiology and Infection* 122 (1999): 7–13.

Bonilla, T. D., K. Nowosielski, M. Cuvelier, A. Hartz, M. Green, N. Esiobu, D. S. McCorquodale, J. M. Fleisher, and A. Rogerson. "Prevalence and Distribution of Fecal Indicator Organisms in South Florida Beach Sand and Preliminary Assessment of Health Effects Associated with Beach Sand Exposure." *Marine Pollution Bulletin* 54, no. 9 (2007): 1472–82.

Bonin, P. "Gonzales Angler Survives Vibrio Infection: Grand Isle Trip Ends in Two-Week Hospital Stay, Ongoing Wound Care." *Louisiana Sportsman*, July 15, 2013. http://www.louisianasportsman.com/details.php?id=5363.

Carroll, L. "Beachgoers Beware: Stomach Bugs Lurk in Sand." NBCNews.com, July 20, 2009. http://www.msnbc.msn.com/id/31928316/ns/health-infectious _diseases.

Centers for Disease Control and Prevention. "*Vibrio vulnificus*." Last updated October 21, 2013. http://www.cdc.gov/vibrio/vibriov.html.

Cipro, C. V. Z., P. Bustamante, S. Taniguchi, and R. C. Montone. "Persistent Organic Pollutants and Stable Isotopes in Pinnipeds from King George Island, Antarctica." *Marine Pollution Bulletin* 64, no. 12 (2012): 2650–55.

Cohen, H. "Dangers at the Beach." *Natural Health Blog*, Baseline of Health Foundation, July 7, 2012. http://www.jonbarron.org/natural-health/raw-sewage -increases-beach-pollution.

Cole, D., S. C. Long, and M. D. Sobsey. "Evaluation of F+ RNA and DNA Coliphages as Source-Specific Indicators of Fecal Contamination in Surface Waters." *Applied Environmental Microbiology* 68, no. 11 (2003): 6507–14.

Converse, R. R., J. L. Kinzelman, E. A. Sams, E. Hudgens, A. P. Dufour, H. Ryu, J. W. Santo-Domingo, C. A. Kelty, O. C. Shanks, S. D. Siefring, R. A. Haugland, and T. J. Wade. "Dramatic Improvements in Beach Water Quality Following Gull Removal." *Environmental Science and Technology* 46 (2012): 10206–13.

Durando, P., F. Ansaldi, P. Oreste, P. Moscatelli, L. Marensi, C. Grillo, R. Gasparini, and G. Icardi. "*Ostreopsis ovata* and Human Health: Epidemiological and Clinical Features of Respiratory Syndrome Outbreaks from a Two-year Syndromic Surveillance, 2005–6, in North-West Italy." *Eurosurveillance* 12, no. 23 (2007). http://www.eurosurveillance.org/ViewArticle.aspx?ArticleId=3212.

Efstratiou, M. A., and A. Velegraki. "Recovery of Melanized Yeasts from Eastern Mediterranean Beach Sand Associated with the Prevailing Geochemical and Marine Flora Patterns." *Medical Mycology* 48 (2010): 413–15.

Elmanama, A. A., M. I. Fahd, S. Afifi, S. Abdallah, and S. Bahr. "Microbiological Beach Sand Quality in Gaza Strip in Comparison to Seawater Quality." *Environmental Research* 99, no. 1 (September 2005): 1–10.

Ernst, E. "Ernst: Rats Blamed for Venice Beach Water Pollution." *Herald Tribune* (Sarasota), May 8, 2012. http://www.heraldtribune.com/article/20120508 /COLUMNIST/120509609.

Esterre, P., and F. Agis. "Beach Sand Nematodes in Guadeloupe: Associated Public Health Problems." *Bulletin of the Society of Pathology and Exotic Filiales* 78, no. 1 (1985): 71–78.

Fellows, J. M., D. No, and M. C. Roberts. "Vancomycin-Resistant *Enterococcus sp.* (VRE) and Methicillin-Resistant *Staphylococcus aureus* (MRSA) in Marine Sand and Water." Poster, Department of Environmental and Occupational Health Sciences, University of Washington, Seattle, n.d.

Fleisher, J. M., L. E. Fleming, H. M. Solo-Gabriele, J. K. Kish, C. D. Sinigalliano, L. Plano, S. M. Elmir, J. D. Wang, K. Withum, T. Shibata, et al. "The BEACHES Study: Health Effects and Exposures from Non-point Source Microbial Contaminants in Subtropical Recreational Marine Waters." *International Journal of Epidemiology* 39 (2010): 1291–98.

Fleming, L. E. "Paralytic Shellfish Poisoning." Woods Hole Oceanographic Institution, 2007. http://www.whoi.edu/science/B/redtide/illness/psp.html.

Food Protection Program. "Red Tide Fact Sheet." Department of Public Health, Executive Office of Health and Human Services, Massachusetts, 2012. http:// www.mass.gov/eohhs/gov/departments/dph/programs/environmental-health /food-safety/red-tide-fact-sheet.html.

Galgani, F., K. Ellerbrake, E. Fries, and C. Goreux. "Marine Pollution: Let Us Not Forget Beach Sand." *Environmental Sciences Europe* 23 (2011): 40–46.

Gast, R. J., L. Gorrell, B. Raubenheimer, and S. Elgar. "Impact of Erosion and Accretion on the Distribution of Enterococci in Beach Sands." *Continental Shelf Research* 31 (2011): 1457–61.

Genuardi, S. "Warm-Water Ocean Bacteria Can Be Life-Threatening." *Sun Sentinel* (Fort Lauderdale), July 23, 2010. http://articles.sun-sentinel.com/keyword /vibrio.

Goodwin, K. D., M. McNay, Y. Cao, D. Ebentier, M. Madison, and J. F. Griffith. "A Multi-beach Study of *Staphylococcus aureus*, MRSA, and Enterococci in Seawater and Beach Sand." *Water Research* 46, no. 13 (2012): 4195–207.

Greig, A. "'Flesh-eating Bacteria' Kills One in Louisiana; Three Others Also Affected." *Daily Mail*, July 7, 2013. http://www.dailymail.co.uk/news/article -2357948/Flesh-eating-bacteria-kills-Louisiana-affected.html#ixzz2YRmlevMS.

Griesbach, A., M. Grimmer, and K. James. *2011–2012 Annual Beach Report Card*. Heal the Bay, Santa Monica, CA, 2012. http://brc.healthebay.org/assets /pdfdocs/brc/annual/2012/HtB_BRC_Annual_2012_Report.pdf.

Groves, M. "Septic Tanks on Their Way Out in Malibu." *Los Angeles Times*, November 6, 2009.

Halliday, E., and R. J. Gast. "Bacteria in Beach Sands: An Emerging Challenge in Protecting Coastal Water Quality and Bather Health." *Environmental Science and Technology* 45 (2011): 370–79.

Hartz, A., M. Cuvelier, K. Nowosielski, T. D. Bonilla, M. Green, N. Esiobu, D. S. McCorquodale, and A. Rogerson. "Survival Potential of *Escherichia coli* and Enterococci in Subtropical Beach Sand: Implications for Water Quality Managers." *Journal of Environmental Quality* 37 (2008): 898–905.

Heaney, C. D., E. Sams, A. P. Dufour, K. P. Brenner, R. A. Haugland, E. Chern, S. Wing, S. Marshall, D. C. Love, M. Serre, R. Noble, and T. J. Wade. "Fecal Indicators in Sand, Sand Contact, and Risk of Enteric Illness among Beachgoers." *Epidemiology* 23, no. 1 (January 2012): 95–106.

Heaney, C. D., E. Sams, S. Wing, S. Marshall, K. Brenner, A. P. Dufour, and T. J. Wade. "Contact with Beach Sand among Beachgoers and Risk of Illness." *American Journal of Epidemiology* 170, no. 2 (2009): 164–72.

Kueh, C. S. W., T.-Y. Tam, T. Lee, S. L. Wong, O. L. Lloyd, I. T. S. Yu, T. W. Wong, J. S. Tam, and D. C. J. Bassett. "Epidemiological Study of Swimming-Associated Illnesses Relating to Bathing-Beach Water Quality." *Water Science and Technology* 31, nos. 5–6 (1995): 1–4.

Lafsky, M. "Fun in the Sand Now Hindered by Fecal Bacteria." *Discoblog* (*Discover* magazine blog), May 14, 2008. http://blogs.discovermagazine.com /discoblog/2008/05/14/fun-in-the-sand-now-hindered-by-fecal-bacteria /#.UcxXMeuXJ_k.

Lee, C. M., T. Y. Lin, C-C. Lin, G. A. Kohbodi, A. Bhatt, R. Lee, and J. A. Jay. "Persistence of Fecal Indicator Bacteria in Santa Monica Bay Beach Sediments." *Water Research* 40, no. 14 (2006): 2593–602.

Levin-Edens, E., O. O. Soge, D. No, A. Stiffarm, J. S. Meschke, and M. C. Roberts. "Methicillin-Resistant *Staphylococcus aureus* from Northwest Marine and Fresh Water Recreational Beaches." FEMS *Microbiology Ecology* 79 (2012): 412–20.

Loureiro, S. T. A., M. A. de Queiroz Cavalcanti, R. P. Neves, and J. Z. de Oliveira Passavante. "Yeasts Isolated from Sand and Sea Water in Beaches of Olinda, Pernambuco State, Brazil." *Brazilian Journal of Microbiology* 36 (2005): 333–37.

Lubick, N. "Dogs Keep Beaches Microbe-free." ScienceNOW, August 31, 2012. *News and Observer* (Raleigh), September 9, 2012.

Lush, T. "31 in Florida Infected by Bacteria in Salt Water." Associated Press, October 11, 2013.

Mozingo, J. "Water Officials Link Malibu Septic Tanks to Beach Pollution." *Los Angeles Times*, November 13, 2000.

Natural Resources Defense Council. "The Impacts of Beach Pollution." In *Testing the Waters*, annual report, 2010, 20–31. www.nrdc.org/water/oceans/ttw /chap2.pdf.

Natural Resources Defense Council. "US Beaches Laden with Sewage, Bacteria: Study." Phys.org, June 27, 2012. http://phys.org/news/2012-06-beaches-laden -sewage-bacteria.html.

Natural Resources Defense Council. *Testing the Waters*, annual report, 2012.

Oshiro, R., and R. Fujioka. "Sand, Soil, and Pigeon Droppings: Sources of Indicator Bacteria in the Waters of Hanauma Bay, Oahu, Hawaii." *Water Science and Technology* 31, nos. 5–6 (1995): 251–54.

Phillips, M. C., H. M. Solo-Gabriele, A. M. Piggot, J. S. Klaus, and Y. Zhang. "Relationships between Sand and Water Quality at Recreational Beaches." *Water Research* 45 (2011): 6763–69.

Plano, L. R., T. Shibata, A. C. Garza, J. Kish, J. M. Fleisher, C. D. Sinigalliano, M. L. Gidley, K. Withum, S. M. Elmir, S. Hower, et al. "Human-Associated Methicillin-Resistant *Staphylococcus aureus* from a Subtropical Recreational Marine Beach." *Microbial Ecology* 65, no. 4 (May 2013): 1039–51.

Ribeiro, E. N., A. Banhos dos Santos, R. F. Gonçalves, and S. T. A. Cassini. "Recreational Water and Sand Sanitary Indicators of Camburi Beach, Vitoria, Es., Brazil." Paper presented at the XXVIII Congreso Interamericano de Ingenieria Sanitaria y Ambiental, Cancún, Mexico, October 27–31, 2002.

Roach, J. "Beach Bacteria Warning: That Sand May Be Contaminated." *National Geographic News*, July 26, 2005. http://news.nationalgeographic.com/news/2005/07/0729_050729_beachsand.html.

Roberts, M. C., O. O. Soge, M. A. Giardino, E. Mazengia, G. Ma, and J. S. Meschke. "Vancomycin-Resistant *Enterococcus* spp. in Marine Environments from the West Coast of the USA." *Journal of Applied Microbiology* 107 (2009): 300–307.

Rock, G. "Is Your Beach Contaminated with MRSA?" *Los Angeles Times*, September 12, 2009.

Sabino, R., C. Verissimo, M. A. Cunha, B. Wergikoski, F. C. Ferreira, R. Rodrigues, H. Parada, L. Falcão, L. Rosado, C. Pinheiro, E. Paixão, and J. Brandão. "Pathogenic Fungi: An Unacknowledged Risk at Coastal Resorts? New Insights on Microbiological Sand Quality in Portugal." *Marine Pollution Bulletin* 62 (2011): 1506–11.

Schrope, M. "Oceanography: Red Tide Rising." *Nature* 452 (2008): 24–26.

Segall, K. "Study: Sticking to the Sand Might Not Be Such Good, Clean Fun for Beachgoers." Stanford News Service, 2007. http://news.stanford.edu/pr/2007/pr-sand-080807.html.

Seriki, D. "Flesh-Eating Beach Bacteria Found in Florida—Death Toll Rises to 9." *SciCraze*, October 1, 2013. http://scicraze.com/2013/10/01/flesh-eating-beach-bacteria-found-florida-death-toll-rises-9/.

Shah, A. H., A. M. Abdelzaher, M. Phillips, R. Hernandez, H. M. Solo-Gabriele, J. Kish, G. Scorzetti, J. W. Fell, M. R. Diaz, T. M. Scott, et al. "Indicator Microbes Correlate with Pathogenic Bacteria, Yeasts and Helminthes in Sand at a Subtropical Recreational Beach Site." *Journal of Applied Microbiology* 110 (2011): 1571–83.

Shibata, T., and H. M. Solo-Gabriele. "Quantitative Microbial Risk Assessment of Human Illness from Exposure to Marine Beach Sand." *Environmental Science and Technology* 46, no. 5 (2012): 2799–805.

Skrzypek, J. "Vibrio vulnificus: Flesh-Eating Ocean Bacteria Hospitalizes 32, Kills 10 in Florida: State Health Department Is Monitoring Bacteria." WPTV.com, October 14, 2013. http://www.wptv.com/dpp/news/region_c_palm_beach _county/palm_beach/vibrio-vulnificus-flesh-eating-ocean-bacteria-hospitalizes -32-kills-10-in-florida.

Soge, O. O., J. S. Meschke, D. B. No, and M. C. Roberts. "Characterization of Methicillin-Resistant *Staphylococcus aureus* (MRSA) and Methicillin-Resistant Coagulase-Negative *Staphylococcus* Spp. (MRCoNS) Isolated from West Coast Public Marine Beaches." *Journal of Antimicrobial Chemotherapy* 64 (2009): 1148–55.

Steele, C. W. "Fungus Populations in Marine Waters and Coastal Sands of the Hawaiian, Line, and Phoenix Islands." *Pacific Science* 2, no. 3 (1967): 317–31. http://hdl.handle.net/10125/7406.

UCLA. "High Levels of Unhealthy Bacteria Found in Sand at L.A. Area Beaches." *Scientific Frontline*, May 23, 2006. http://www.sflorg.com/sciencenews /scn052306_02.html.

UCLA. "Study Shows Unhealthy Bacteria in Southern California Beach Sand." Phys.org, May 23, 2006. http://phys.org/news67624127.html.

University of North Carolina, Chapel Hill. "Study: Digging in Beach Sand Increases Risk of Gastrointestinal Illness." Phys.org, July 9, 2009. http://phys .org/news166372659.html#nR1v.

U.S. Environmental Protection Agency. "Health and Environmental Effects Research: Digging in Beach Sand Linked to Increased Risk of Gastrointestinal Illness." 2012. http://www.epa.gov/nheerl/articles/2012/Digging_in_beach_sand .html.

Valdes-Collazo, L., A. J. Schultz, and T. C. Hazens. "Survival of *Candida albicans* in Tropical Marine and Fresh Waters." *Applied Environmental Microbiology* 53, no. 8 (1987): 1762–67.

Vogel, C., A. Rogerson, S. Schatz, H. Laubach, A. Tallman, and J. Fell. "Prevalence of Yeasts in Beach Sand at Three Bathing Beaches in South Florida." *Water Research* 41, no. 9 (2007): 1915–20.

Vorsino, M. "Group Testing Waikiki Sand for Bacteria." *Honolulu Star Bulletin*, April 20, 2006. http://archives.starbulletin.com/2006/04/20/news/story07.html.

Whitman, R. L. "Beach Sand Often More Contaminated Than Water." USGS Newsroom, September 12, 2008. http://www.usgs.gov/newsroom/article.asp?ID=2022.

Whitman, R. L., and M. B. Nevers. "Foreshore Sand as a Source of *Escherichia coli* in Nearshore Water of a Lake Michigan Beach." *Applied Environmental Microbiology* 69, no. 9 (2003): 5555–62.

Whitman, R. L., D. A. Shively, H. Pawlik, M. B. Nevers, and M. N. Byappanahalli. "Occurrence of *Escherichia coli* and Enterococci in *Cladophora* (Chlorophyta) in Nearshore Water and Beach Sand of Lake Michigan." *Applied Environmental Microbiology* 69, no. 8 (2003): 4714–19.

Winner, C. "Shifting Sands and Bacteria on the Beach: Does Ever-Moving Sand

Transport Microbes along with It?" *Oceanus*, September 1, 2011. https://www
.whoi.edu/oceanus/viewArticle.do?id=110889.

Woods Hole Oceanographic Institution. "Beach Closures." Accessed September
10, 2012. http://www.whoi/edu/main/topic/beach-closures.

Yamahara, K. M., B. A. Layton, A. E. Santoro, and A. B. Boehm. "Beach Sands
along the California Coast Are Diffuse Sources of Fecal Bacteria to Coastal
Waters." *Environmental Science and Technology* 41, no. 13 (2007): 4515–21.

Yamahara, K. M., S. P. Walters, and A. B. Boehm. "Growth of Enterococci in Un-
altered, Unseeded Beach Sands Subjected to Tidal Wetting." *Applied Environmental Microbiology* 75, no. 6 (2009): 1517–24.

Zieger, U., H. Trelease, N. Winkler, V. Mathew, and R. N. Sharma. "Bacterial
Contamination of Leatherback Turtle (*Dermochelys coriacea*) Eggs and Sand
in Nesting Chambers at Levera Beach, Grenada, West Indies—a Preliminary
Study." *West Indian Veterinary Journal* 9, no. 2 (2009): 21–26.

Chapter 9. The International Dimension of Beach Destruction

Adam, D., and J. Vidal. "Britain Accused of 'Double Counting' over Climate
Aid to Bangladesh." *Guardian*, July 13, 2009. http://www.guardian.co.uk
/environment/2009/jul/13/climate-change-development.

Esmaquel, P., II. "Cebu Launches Crackdown on Illegal Seashell Trade." GMA
News Online, May 18, 2011. http://www.gmanetwork.com/news/story/220937
/news/specialreports/cebu-launches-crackdown-on-illegal-seashell-trade.

European Commission. "Guyana." Development and Cooperation—Europeaid,
last updated February 17, 2012. http://ec.europa.eu/europeaid/where/acp
/country-cooperation/guyana/guyana_en.htm.

Gornitz, V. *Rising Seas: Past, Present, Future.* New York: Columbia University
Press, 2013.

Macauhub News Agency. "Mozambique: World Bank Funds Building Work to
Stop Coastal Erosion in Town of Vilanculos." April 22, 2008. http://www
.macauhub.com.mo/en/2008/04/22/4916/.

Macauhub News Agency. "Saudi Arabia to Grant Mozambique Loan for Coastal
Protection of City of Maputo." May 26, 2011. http://www.macauhub.com
.mo/en/2011/05/26/saudi-arabia-to-grant-mozambique-loan-for-coastal
-protection-of-city-of-maputo/.

Pilkey, O. H., and M. E. Fraser. *A Celebration of the World's Barrier Islands.* New
York: Columbia University Press, 2003.

Pilkey, O. H., W. J. Neal, J. T. Kelley, and J. A. G. Cooper. *The World's Beaches:
A Global Guide to the Science of the Shoreline.* Berkeley: University of California Press, 2011.

Pilkey, O. H., and K. C. Pilkey. *Global Climate Change: A Primer.* Durham, NC:
Duke University Press, 2011.

Pilkey, O. H., and R. Young. *The Rising Sea.* Washington, DC: Island Press, 2009.

Queface, A. *Climate Change Impacts and Disaster Risk Reduction in Mozambique.*

Instituto Nacional de Gestão de Calamidades, Ministério da Administração Estatal, September 14, 2012. http://www.sarva.org.za/sadc/download/moz2012 _10.pdf.

Chapter 10. The End Is Here

Boretti, A. A. "Discussion of J. A. G. Cooper, C. Lemckert, Extreme Sea-Level Rise and Adaptation Options for Coastal Resort Cities: A Qualitative Assessment from the Gold Coast, Australia." *Ocean and Coastal Management* 78 (2013): 132–35.

Cooper, J. A. G., and C. Lemckert. "Response to Discussion by A. Boretti of Cooper, J. A. G., and Lemckert, C., Extreme Sea-Level Rise and Adaptation Options for Coastal Resort Cities: A Qualitative Assessment from the Gold Coast, Australia, *Ocean and Coastal Management* 64 (2012): 1–14." *Ocean and Coastal Management* 78 (2013): 136–37.

Cooper, J. A. G., and C. Lemckert. "Extreme Sea-Level Rise and Adaptation Options for Coastal Resort Cities: A Qualitative Assessment from the Gold Coast, Australia." *Ocean and Coastal Management* 64 (2012): 1–14.

Gornitz, V. *Rising Seas: Past, Present, Future.* New York: Columbia University Press, 2013.

Greenwood, A. "Rising Sea Levels Leave National Trust Properties in Devon and Cornwall at Risk." *Western Morning News*, January 18, 2014. http://www .westernmorningnews.co.uk/Rising-sea-levels-leave-National-Trust-properties -in-Devon-and-Cornwall-at-risk.

National Trust. *Shifting Shores. Living with a Changing Coastline.* Annual report, 2005.

Pilkey, O. H., and M. E. Fraser. *A Celebration of the World's Barrier Islands.* New York: Columbia University Press, 2003.

Pilkey, O. H., W. J. Neal, J. T. Kelley, and J. A. G. Cooper. *The World's Beaches: A Global Guide to the Science of the Shoreline.* Berkeley: University of California Press, 2011.

Pilkey, O. H., and K. C. Pilkey. *Global Climate Change: A Primer.* Durham, NC: Duke University Press, 2011.

Pilkey, O. H., and R. Young. *The Rising Sea.* Washington, DC: Island Press, 2009.

Sheppard, T. *The Lost Towns of the Yorkshire Coast.* London: A. Brown and Sons, 1912.

University of Ulster. "Iconic Beach Resorts May Not Survive Sea Level Rises." *ScienceDaily*, January 16, 2013. Accessed September 30, 2013. http://www .sciencedaily.com/releases/2013/01/130116090642.htm.

Woodroffe, C. D. *Coasts: Form, Process, and Evolution.* Cambridge, UK: Cambridge University Press, 2002.

Worrall, S. "The UK LIFE Project on Shoreline Management: 'Living with the Sea.'" In "Proceedings 'Dunes and Estuaries 2005'—International Conference on Nature Restoration Practices in European Coastal Habitats, Koksijde, Belgium, 19–23 September 2005." *VLIZ Special Publication* 19 (2005): 451–59.